DESIGN LIKE YOU REALLY MEAN IT

Design Like You Really Mean It

Mean It

HEIDI BROWN

To so many people in my life who listen to my conversations on how things work in physical and digital form and the thinking and decisions that were to make that experience positive or negative. This stuff is my life, my passion. Best rant ears are my family. My husband and kids. Thank you, Mr. Brown for your unwavering support and curiosity in me. And my two daughters, you help me be me. I have grown so much in seeing the world through your eyes in every stage; my curiosity and intention muscles have expanded; that is an amazing gift you've given me.

CONTENTS

Author's Note

I got stuck! What did I do? What I normally do…

Bring intention. This is one of my favorite words in design. Having intention and thought around where you're headed, intention around what experience you are creating and intention around how you're making it. Now I'm using that word for myself, the intention of a book. Intention on giving back to a community I love.

Being a designer is the most amazing job; it's also very taxing. I'm so grateful for so many people in my life who encouraged me, who helped me through tough times, who listened.

Now it's my turn to pass this on. I want to give this community tools from learnings I've had as a designer on a team as well as leadership roles across different industries. I'm offering these lessons, so you have something to build from.

On Using AI in This Book

In the creation of this book, I used AI as a supportive tool, not a source of ideas. Every insight, story, and point of view comes from my own experience, thinking, and lived practice. The first draft of this book was written entirely by me, directly from my own mind and work. Later in the process, I used AI selectively to help organize complex thoughts, reduce repetition, check grammar, and assist with light editing for clarity. Throughout, I applied my own discernment, choosing when to accept a suggestion, when to reject it, and when my own judgment better served the point I was trying to make. Much like a thoughtful editor, AI helped surface patterns and tighten language, but the substance, perspective, and intent remained fully mine. This book reflects my beliefs, my judgment, and my voice. AI helped with the polish, but the thinking and decisions behind it are wholly my own.

Introduction

Hello designer friends,

This is not just a book about design. It's about *becoming* the kind of designer who designs like they mean it. Whether you're mid-level and seeking more voice, or leading others through complexity and change, this book is your invitation to level up, not just in skill, but in self-awareness, influence, and creative conviction. Because to design like you really mean it is to:

- Show up fully, not just functionally
- Connect strategy to soul
- And own the outcomes your work creates, beyond the screen

You'll move through chapters that challenge your thinking, expand your toolkit, and stretch your creative identity. Along the way, you'll engage in deep reflection and visual exploration that will help you uncover your **Soulprint**; the evolving blueprint of who you are as a designer when you're most aligned, intentional, and alive in your craft. Inside, you'll find:

- A roadmap from execution to strategic influence
- Tools to navigate tension, ambiguity, and impact
- Prompts to clarify your purpose and power
- Artistic exercises that unlock your intuition
- A final Soulprint, your personalized, visual artifact of growth

Design Like You Really Mean It is more than a rallying cry. It's a call to action — and a compass for designing your future. One decision, one insight, one Soulprint at a time.

How To Use This Book (And Make It Yours)

This is not just a book to read.

It's a book to work through, reflect with, and create alongside.

A companion for your inner and outer transformation as a designer.

Each chapter offers not just insight, but an invitation. To pause. To look inward. To notice how your values, experiences, and instincts are shaping your work. And to begin honoring those signals with clarity and courage.

As you move through the chapters, you'll find three special elements designed to deepen your journey:

Reflection Prompts

At the end of each chapter, you'll discover a set of prompts to help you process what you've read, not just intellectually, but emotionally and experientially. These are designed to help you surface truths, clarify blind spots, and reconnect with the designer you're becoming.

Some prompts will feel natural. Others may challenge you. That's the point. Growth requires curiosity, and you're here to grow.

Artistic Expression

Designers are makers. And sometimes the most honest insights can't be put into words, they need to be drawn, sketched, mapped, or shaped.

Each chapter includes an artistic prompt: a visual metaphor, collage idea, or creative exercise that helps you give form to what's shifting inside you.

You don't need to be "an artist." You just need to be open. Let yourself play. Use scraps, stick figures, color bursts, symbols. This isn't about polish; it's about your soul.

Soulprint Words

As you reflect, you'll be asked to name a few words that represent the essence of what you're uncovering. These words become the threads of your *Soulprint*.

Your Soulprint is the living, breathing artifact of who you are as a designer: the values that guide you, the tensions you hold, the aspirations you carry, and the superpowers you're starting to own.

The Final Chapter: Create Your Soulprint

At the end of this book, you'll be invited to bring everything together.

Using the words, symbols, and sketches from each chapter, you'll create your own *Designer Soulprint*: a visual representation of the designer within you — who you are when you're most aligned, most courageous, and most alive in your work.

Whether it becomes a digital collage, a drawing, a mood board, a journal spread, or something totally unexpected, it will be yours.

A reminder. A compass. A declaration.

| 1 |

Turning Business into the Human Experience

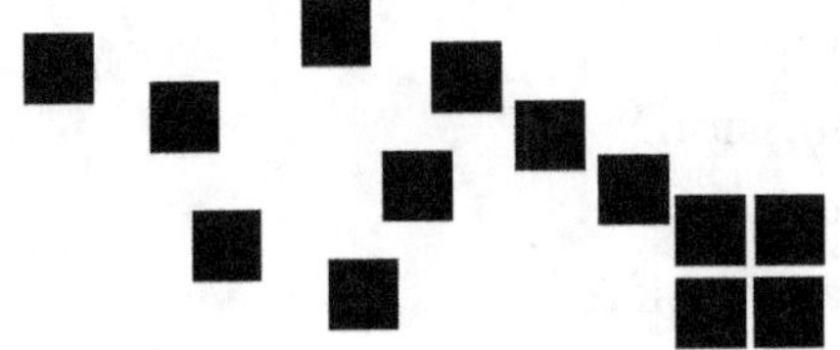

What the Heck Is Design?

Let's start wide. Because design isn't just one thing, it's everything and something very specific all at once. It shows up in any context, any company, any industry. Design is the invisible thread between a problem and its solution, and yet it's not just the solution. It's how we move toward the solution. It's how the end user feels about the solution. It's how the team arrives there together.

Design is the process of mapping ambiguity to clarity. It's naming what's unsaid, shaping it, and then breathing life into it.

And when done well, design can change the course of a business, from floundering to thriving. I've seen it happen. I've been part of it. I've watched design revive tired products, align teams in chaos, and give breath to brand-new ideas struggling to stand on their own. And yes, sometimes it doesn't work. Because design isn't free. It takes time. Talent. Thought. And collaboration. It takes people who can listen, make sense of patterns, and then work with engineers, product leaders, and marketers to turn a sketch of possibility into something real.

So, what is design… really?

Design is how something feels, yes. But also, how it functions.

Design is process.

Design is storytelling.

Design is structure.

Design is texture.

Design is expression.

Design is emotion.

Design is data, translated.

Design is a bridge between what is and what could be.

Design is how it works and why it matters.

It's a conversation between intention and reality, between what a designer puts into the world, and what the world feels, thinks and does in response.

Design in the Modern World

Today, design starts with what's visible: the brand, the interface, the layout, the typography, the interactions. But it goes deeper, into how it all comes together to serve a user need, tell a story, or empower someone to do something they couldn't before.

You may be known in your organization as "the visuals person", the one who makes mockups, prototypes, social assets, or the product UI. And yes, those are expressions of design. But you're not just making things look good; you're making them make sense. You're giving form to a strategy. You're making decisions about emotion, hierarchy, communication, access, equity.

Design is not just aesthetics. It's alignment.

It's not just polish. It's purpose.

It's not just the surface. It's the system.

We're entering an era of personalization, where experiences must meet people exactly where they are. And that means we're entering the era of design as connective tissue, between product and people, data and action, value and meaning.

Design doesn't live in the pixels; it lives in the outcomes.

So, what is design?

Design is your way of shaping possibility.

Design is your voice in the room when strategy is on the table.

Design is your power to connect what's human with what's built.

And you are not just a designer.

You are a connector. A translator. A systems thinker.

You are a leader, whether your title says so yet or not.

Let's design like we mean it.

Design to Drive Business

When you embrace the design process and craft, design is truly an innovation driver. If you think about all the big companies with name recognition, they all have mature design organizations and they know it is a competitive advantage. These organizations understand that design is the center of superior customer experiences that drive loyalty and repeat business because it is centered on users at the heart.

Organizations can create value with design in two ways. One is by designing value for customers, and the second is by optimizing organizational efficiency; and magic happens when there is external and internal alignment. Design can foster a design-centric culture within an organization. This could include hiring practices, training programs, and leadership support.

Design can reduce costs, increase revenue, grow a business, open new opportunities, open new markets, increase efficiency, extend reach, mitigate risk and amplify word of mouth.

Design plays a critical role in business by bridging the gap between functionality and user experience, enhancing both product appeal and market competitiveness. Design is the secret sauce! It is what takes the boring and makes it something memorable.

Design is responsible for leading cross-functional collaboration, with a group and outside of it. Meaning, inside product, design and engineering culture or inside brand but then between brand and product and even sales. Design can also be at the spear leading agile and lean methodologies, which promotes iterative development and continuous improvement.

Process of Design

Design isn't magic, even if it feels that way sometimes. Behind every moment of clarity is a structure, a rhythm, that helps us move from chaos to coherence. This is where the process of design comes in.

There are two aspects of design. There's the craft of design, the tangible expression of creativity through visuals, systems, and stories. And then there's the process of design, the mindset and journey that can be applied to nearly any challenge a business faces.

The process isn't meant to box creativity in. It's there to channel it. Think of it as the language design speaks when it translates complexity into meaning. When you understand the process, you gain the ability to connect the right tool with the right need, to know when to explore, when to decide, when to slow down, and when to build momentum.

Your challenge is to find your own way of moving through it. You may lead with empathy, lean toward systems thinking, or find your strength in storytelling. No matter your style, every project travels through familiar territories: moments of understanding, ideation, creation, testing, and reflection.

Good design reaches into deep corners of the brain, understanding not just behavior but feeling and emotion. Sometimes the path is messy and nonlinear. Other times it's sharp and structured. The shape of the journey depends on what you're designing and what outcome you're aiming for.

What follows can look overwhelming at first. That's intentional, not because you're meant to do all of it, but because it shows the range of how design thinking can show up. This isn't a checklist or a prescription. It's a way to build awareness.

Every element below is a point in the design process. Every project is different. The time you spend in each area will expand or contract depending on context, constraints, and goals. Some of these activities you already do instinctively. Others may feel unfamiliar or uncomfortable.

As you read through them, notice what you naturally gravitate toward and what you tend to skip or rush. Pay attention not just to *what* you do, but *how* you do it. With how much intention? With what level of depth or precision? Are you exploring to truly learn, or just to confirm what you already believe?

Understanding your tendencies isn't about judgment. It's about choice. When you know where you thrive and where you struggle, you can design your own balance, leaning into your strengths while deliberately stretching into new territory when the work calls for it.

What follows isn't a formula. It's a map of the design journey, a set of mindsets, methods, and activities that guide how you think, make, and collaborate. Consider it your design toolset and a lens you can return to throughout this book, not to do more, but to design more consciously.

DESIGN IS THE THREAD

Empathize:

- Understand: Gain insight into users' needs, behaviors, and motivations.
- Observe: Watch users in their natural environment to gather information.
- Immerse: Engage deeply with users to experience their challenges first-hand.
- Engage: Interact directly with users through interviews and conversations.

Define:

- Clarify: Make the problem and user needs clear and precise.
- Specify: Detail the problem scope and constraints.
- Frame: Establish a perspective or point of view on the problem.
- Articulate: Clearly express the problem statement and user needs.

Ideate:

- Brainstorm: Generate a large number of ideas without judgment.
- Generate: Produce various concepts and solutions.
- Conceive: Formulate new ideas and innovative solutions.
- Explore: Investigate different possibilities and alternatives.

Prototype:

- Create: Build tangible representations of ideas.
- Build: Develop functional models or simulations.

- Construct: Assemble prototypes using available materials and tools.
- Model: Design representations that can be tested and refined.

Test:

- Validate: Ensure the solutions meet user needs and solve the problem.
- Evaluate: Assess the effectiveness and usability of the prototypes.
- Assess: Measure the success and impact of the solutions.
- Experiment: Try out prototypes in real-world scenarios to gather feedback.

Additional activities that can be applied to any part of the process:

Research:

- Investigate: Conduct detailed studies to gather information.
- Analyze: Break down information to understand patterns and insights.
- Study: Examine in detail to understand various aspects.
- Inquire: Seek information through questioning and exploration.

Synthesize:

- Integrate: Combine different pieces of information into a coherent whole.
- Collate: Collect and organize information systematically.
- Summarize: Condense information into key points.

- Distill: Extract the essential elements from the gathered data.

Collaborate:

- Cooperate: Work together with others to achieve common goals.
- Partner: Form alliances with others to enhance the design process.
- Co-create: Jointly create solutions with stakeholders.
- Engage: Actively involve team members and users in the process.

Iterate:

- Refine: Make continuous improvements based on feedback.
- Improve: Enhance the design to better meet user needs.
- Revise: Make changes and updates to the design.
- Update: Keep the design current with new insights and feedback.

Visualize:

- Illustrate: Create visual representations of ideas and concepts.
- Sketch: Draw rough representations to communicate ideas.
- Depict: Show ideas through visual means.
- Map: Create diagrams to show relationships and flows.

Communicate:

- Present: Share ideas and solutions with stakeholders.

- Share: Disseminate information with the team and users.
- Explain: Make ideas and concepts clear to others.
- Convey: Communicate the essence of ideas effectively.

Facilitate:

- Guide: Lead the team through the design process.
- Moderate: Oversee discussions and activities to ensure productive outcomes.
- Lead: Direct the process and keep it on track.
- Coordinate: Organize activities and resources effectively.

Reflect:

- Review: Look back at the process and outcomes to learn from them.
- Contemplate: Think deeply about the implications and next steps.
- Consider: Take into account various factors and feedback.
- Ponder: Reflect on the insights gained and their meaning.

Specific Techniques or Activities:

- Storyboard or user journey mapping: Plan the sequence of events or interactions.
- Map: Chart user journeys, processes, or information architecture.
- Pilot: Conduct small-scale tests to validate concepts.
- Trial: Run experimental tests to gather data.
- Demo: Show prototypes to stakeholders for feedback.
- Prove: Demonstrate the feasibility and effectiveness of solutions.

- Prioritize: Rank ideas and solutions based on importance and feasibility.
- Rank: Order items by criteria.
- Order: Arrange elements systematically.
- Assess: Evaluate options and make decisions.
- Select: Choose the best options from alternatives.
- Cluster: Group similar items together.
- Categorize: Organize items into categories.
- Organize: Arrange items in a structured way.
- Classify: Sort items into predefined groups.
- Mock-up: Create a visual or functional representation.
- Draft: Create initial versions for feedback.
- Sample: Produce examples for testing.
- Simulate: Mimic real-world scenarios to test solutions.

From Seat to Stage

When you finally find yourself in that coveted room where strategy is discussed and decisions are made, don't shrink. Don't wait to be asked. You're not just there to listen. You're there to shift perspective. To spark new ways of seeing.

I've seen designers, including myself, go quiet in those moments. But I've also learned this: for design to be valued, it has to be felt. And to be felt, it has to be heard. So, speak. Ask generous, curious questions. That's a superpower of design. Use your ability to see around corners and across systems to reveal things the business hasn't yet named. Help people understand the problem before they rush to solve it. That's what design is for.

Designers are more than execution partners. You are a thought partner.

We are systems thinkers, story shapers, pattern identifiers. But that power only matters if we learn how to communicate it clearly.

Honestly, I struggle with this daily. My brain connects ideas faster than I can speak. I often worry that if I shared everything happening in my head, it might overwhelm the room. So, I've learned to pace my vision. Sometimes I show the end state and work backward. Other times, I paint it in words.

Visuals are powerful. People anchor on them. So, use them with intention. Know your desired outcome before you drop that beautiful Figma frame or that conceptual sketch. Design can light the path or distract from it, depending on how and when you share it.

Let the business experience design. Not just as an output, but as a way of thinking, feeling, and deciding. And if the room doesn't quite know how to hold space for that yet? Build your own room. Invite others in. That's how design leads.

Carry This Forward

Design isn't decoration; it's translation. It's how strategy becomes something people can feel. Carry forward the awareness that every design decision connects business intention to human experience.

Chapter Reflection

Reflection Prompts

- When have you seen your design work drive true business impact?
- Where do you feel your value is misunderstood, or underestimated?
- What would change if you saw yourself not as a service partner, but a strategic leader?

Artistic Expression

Draw, paint, or collage a "thread of transformation." Begin with a swirl of chaos and slowly pull a line through it — this line represents your impact. Where does it take you? What symbols or patterns emerge as clarity arrives?

Choose 3–5 words that capture what this chapter unlocked in you.

Examples: *Strategic, Connector, Sense-maker, Vital, Amplifier*

Soulprint Thread

This chapter reveals your role as a transformer of ambiguity into clarity.

Carry forward one image, word, or symbol that represents how you create meaning that matters.

| 2 |

Be the Bridge

We've seen how design translates business into experience. Now, let's explore how designers become the bridge between vision and outcome.

How to Connect Design to Business Outcomes Without Losing Your Creative Soul

Great design doesn't begin with pixels; it begins with clarity. That means defining the inputs before the ideas, knowing what success looks like not just from a user's perspective, but from the businesses too.

Design is part of an interconnected ecosystem. It is both imaginative and practical. We need it to push ideas forward and ground those ideas. And designers — *you* — are often the only ones trained to see across that full picture.

Step 1: Envision the End State

Start by imagining the product as if it already exists. What does success feel like for the user? What problem does it solve? What has changed in their behavior or perception? Don't worry about design just yet, stay in the story.

Try writing a mock press release that highlights the launch. Focus on the big ideas: what the product does, what makes it valuable, and why it matters. If you're not a writer, that's okay. Use bullet points or short phrases. Just focus on the outcome, not the interface.

Step 2: Visualize the Vision (But Don't Design It Yet)

Translate that vision into a simple, compelling slide or image. Think about the essence of success — what changes for the user? Maybe it's time saved, mobile adoption, fewer errors, or increased confidence. You're not designing the final product. You're sketching possibility. This can be a diagram, a flow, or a storyboard. Be careful not to jump into UI too soon — visuals are powerful, and stakeholders tend to anchor on them. Use that power wisely.

Step 3: Define Success Metrics

Good design is measurable. Create a few user-centric metrics tied to your imagined end state. Maybe it's task completion rates, NPS, speed to value, or adoption within a specific audience. These metrics act like lighthouses, they truly help everyone steer in the same direction. Without them, design risks becoming subjective or reactive.

As a designer, your job is not only to solve problems, but also to frame them. And you can't do that alone. You need to connect across the organization. Build partnerships with product marketing, marketing, product management, project and program managers, data analysts, and engineers. These partners help shape the landscape. They hold different perspectives on what success means, and your role is to bridge those perspectives into one clear story.

The Surface and the Flow

Here's something I've learned over time: the flow of an experience is a designer's responsibility. But the surface — the platform, medium, or channel — should be a shared decision. Is it an in-app experience? An email? A Slack notification? Should this be part of an existing surface or something new? Design is a very strong influence and should be, however there is much at stake with these decisions and should be shared decisions for the team in order to make the right investment, at the right time.

Things to ask:

- Do we have the infrastructure to deliver this?
- Does it belong in the current user journey, or should we create a new one?
- Will it require building from scratch or repurposing something we already have?

These are product questions. These are design questions. They're also partnership questions. You don't need to know all the answers, but you do need to help facilitate the conversation.

Bridging Begins with Inputs

When you set clear inputs: a vision, a shared understanding of success, and a few key metrics, you give your team something powerful... a target. Not a list of tasks. Not a vague sense of "make it better." A direction. That is how design creates value.

This works especially well on big initiatives, but even small requests can benefit. If someone asks for a social media asset, don't just ask for the dimensions. Ask what outcome they're hoping for. Who's it for? What will success look like? Turn every request into a moment of clarity. Your team will thank you. You'll start designing with more freedom and confidence.

You May Already Be Bridging Without Knowing It

As a designer, you may already be guiding your partners toward clarity without them realizing it. That's leadership. Many partners haven't worked deeply with design before or may not know this flavor of clarity. They may not know how to brief you. They may see you as a service, not a strategy partner. This is your chance to reshape that.

I don't believe in design teams working from laundry lists. That's not design, that's production. And sure, sometimes you'll need to execute. But if you want your work to have impact, it starts with intention. It starts with you being the bridge.

Every organization has a different understanding of what design is and what designers do. That perception will vary between teams, roles, and even individuals. So, ask yourself: how is design seen where you work? Do people understand its value? Do they invite it in early or late?

Be honest about where things stand. That's the first step to shifting the culture, and your role in it.

You are not just here to design screens.

You are here to connect ideas to outcomes.

You are the bridge.

Carry This Forward

When you define success before design begins, you move from execution to influence. Carry forward the mindset that your role is not just to make, but to connect; to bridge clarity between design and business, people and purpose.

Chapter Reflection

Reflection Prompts

- When have you served as a bridge between design and another function?
- What happens when you define success up front? What happens when you don't?
- Where can you advocate more clearly for design's role in shaping outcomes?

Artistic Expression

Sketch or map a bridge across a canyon — label each side (design | business) and illustrate the planks that make it crossable: communication, trust, clarity, intent. What's missing? What's strong?

Soulprint Words

Choose 3–5 words that capture what this chapter unlocked in you.

Examples: Translator, Clarifier, Synthesizer, Navigator, Framer

Soulprint Thread

This chapter reveals your power to *align through vision.*

Carry forward a phrase or image that reminds you: *"I help teams see together."*

Framework for Intuitive Listening

Bridging begins with listening. The next step is learning how to hear beyond words; to navigate complexity through intuition and empathy.

The Inner Compass

Designers often sit at the intersection of information, emotion, and intention. Listening isn't just about hearing users or stakeholders; it's about orienting yourself in complexity and navigating it with purpose. That's why I'm introducing The Compass Framework, a tool for you to become navigators of change, grounded in empathy, guided by curiosity, and aligned with outcomes.

This framework isn't about a linear checklist. It's a directional tool that helps us tune into subtle signals, spot tradeoffs, and move design forward with integrity and intention.

I've spent a great deal of time reflecting on what it truly means to listen and how it shapes an empathetic, intuitive designer. This perspective is deeply rooted in my experiences; how I respond to strategy, engage with my team, and guide designers in showing up fully in their work. I hope this framework resonates with you and I invite you to try it. It's more than a metaphor; it's a way to approach design from a fresh perspective, where rhythm and responsiveness are as vital as strategy and execution.

The Compass Framework

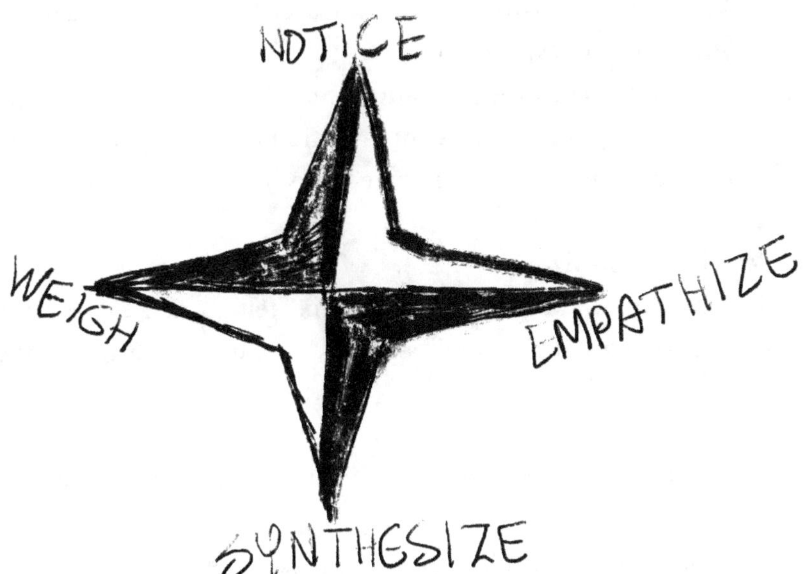

North: Notice

Start by noticing what's happening, without judgment. This is where you bring your awareness to user signals, body language, stakeholder tension, or what isn't being said.

- What are you sensing in the room or interview?
- What patterns are emerging from feedback, even if subtle?
- What's *not* being said that feels important?

East: Empathize

Move with empathy, not just for users, but for the business, your teammates, and yourself. Empathy expands the field of understanding before narrowing toward action.

- Move with empathy, not just for users, but for the business, your teammates, and yourself.
- What are their pressures and incentives?
- Where might fear, hope, or ambition be driving behavior?
- How do you allow space for feelings in decision-making?

South: Synthesize

Sift through what you've noticed and felt to make meaning. You are drawing connections between the human truth and the business needs.

- What's the tension between what people want and what's possible?
- What deeper problem or insight is underneath the feedback?
- What might this look like as a design principle?

West: Weigh

Here's where you pause and reflect. Listen to your gut. Weigh what you know with what you feel.

- Are you moving too fast to respond with empathy?
- Is there a potential tradeoff being made that needs to be surfaced?
- Are you designing for what's most urgent? Or what's most important?

Why a Compass?

Because intuitive design isn't a straight line. It's a navigation of competing forces, data vs. gut, clarity vs. ambiguity, user desire vs. business constraint. You'll move in all directions throughout your process. This framework gives designers a structured but flexible way to build awareness, build trust, and ultimately, build better outcomes.

Intuition is not the opposite of data; it is what helps you hear what data cannot say. While numbers reveal patterns, intuition helps you recognize emotional undercurrents, unmet needs, or emerging tensions. Listening deeply often involves trusting your gut when something doesn't feel right, even if you can't yet explain why. The compass helps you tune in to those signals, then test and validate them in thoughtful, rigorous ways. Design is both analytical and emotional. It asks you to hold space for ambiguity while moving toward clarity.

Balancing Data and Intuition

Listening with intention requires you to move fluidly between gut instinct and external validation. Sometimes your instincts will surface an insight before it shows up in the data. That does not make it wrong, it makes it worth exploring.

Trusting your gut can feel risky in design cultures that prioritize metrics or hard proof. But often, the most resonant ideas begin as a felt sense. The key is to stay grounded in empathy and open to feedback. Let your intuition guide you toward the questions and let feedback help shape the answers.

Compass Exercises

Use the questions below to reflect on how you move through this framework and how your listening, intuition, and decisions show up in service of your business and the people it serves.

Map the Compass

Choose a recent project or conversation. Use the compass to reflect on each direction. What did you notice, where did you empathize, what did you synthesize, and what tradeoffs did you weigh?

Build Your Listening Rituals

What practices help you tune in? Is it journaling after meetings? Is it sketching feelings instead of just solutions?

Spiderweb of Sparks

Identify the moments that sparked an idea or shifted your thinking. Place them along the compass to track your intuitive map.

Your Listening Style

Are you more of a noticer, synthesizer, or empathizer? Where do you want to grow?

Gut Check Practice

Think back to a time when your intuition told you something before the data confirmed it. What did it feel like? How did you act on it? What happened next? Use the compass to walk through that experience. What did you notice? What tradeoff did you weigh? What helped you trust your gut, or ignore it?

Listening as a Creative Act

Great design comes from deep listening. But it is not passive. Listening, when done well, is one of the most active and creative acts a designer can practice. It is about more than understanding others. It is about finding clarity in ambiguity, making space for emotion and instinct, and choosing direction in a moment of uncertainty.

Listening deeply is what allows us to tap into the emotional undercurrents that data often misses. As a designer, you are constantly tuning into signals, many of which are not spoken aloud. Intuition matters. Trusting your gut when something feels off or leaning into an idea that has no clear data trail but feels emotionally right can lead to meaningful breakthroughs.

The compass is meant to support you through all of that. Let it guide you, challenge you, and return you to the heart of what matters most: listening with intention, designing with empathy, and creating with purpose.

In my approach to design, I've never simply done what was asked of me. Instead, I make it a priority to listen deeply and truly understand the underlying problems. By immersing myself in the context and complexities of each challenge, I'm able to identify solutions that genuinely meet the needs at hand. It's not about following instructions, it's about uncovering the real issues and delivering thoughtful, impactful design that addresses those core needs.

You can develop this sensitivity through intentional practice. Engage in empathetic inquiry by walking in the user's shoes, imagining their daily frustrations, and building rapport during conversations. Observe them in their natural environment to uncover unspoken needs. Spend time immersed in their world. This helps you see patterns not just in behavior, but in emotion and intention.

Use empathy maps to capture what users think, feel, say, and do. This creates a multidimensional picture that informs design in powerful ways. Share those stories with your team. Let the voice of the user be part of your collaboration, not just your research.

When analyzing data, look for emotional patterns. Where are users consistently feeling joy, frustration, or hesitation? Cross-reference these feelings with quantitative metrics to paint a richer picture.

Feedback loops matter. Not just the formal kind, but the ongoing, conversational ones. Feedback is where your instincts meet the realities of your team and your users. Sometimes, it confirms what you already feel. Other times, it challenges you to stretch. It keeps your creative compass calibrated.

Designers grow through the ability to integrate emotional intelligence with analytical reasoning. Merging gut, heart, and brain helps you stay grounded. Let your instincts lead when imagining but check them with research. Let your heart drive your storytelling but let your team help shape it. Let your brain refine and systematize what you know intuitively.

This is not about romanticizing intuition. It is about creating a healthy relationship between instinct and evidence, emotion and logic, solo insight and shared critique.

Carry This Forward

Listening is a creative act. Through intuition, empathy, and awareness, you begin to sense the unseen forces shaping your work. Carry forward your compass, your ability to notice, empathize, synthesize, and weigh with intention.

Chapter Reflection

Reflection Prompts

- Which compass direction (notice, empathize, synthesize, weigh) feels strongest for you? Which is least comfortable?
- How do you listen when no one is speaking? What cues do you notice in silence?
- When has your gut been right before the data confirmed it?

Artistic Expression

Create your own compass. Assign colors, textures, or symbols to each direction. Let it reflect how you navigate; where you pause, how you process, and what pulls you toward truth.

Soulprint Words

Choose 3–5 words that capture what this chapter unlocked in you.

Examples: *Attuned, Intuitive, Grounded, Empathetic, Perceptive*

Soulprint Thread

Your internal compass takes shape here. Carry forward a symbol or sketch that reminds you: *"I know how to navigate ambiguity with integrity."*

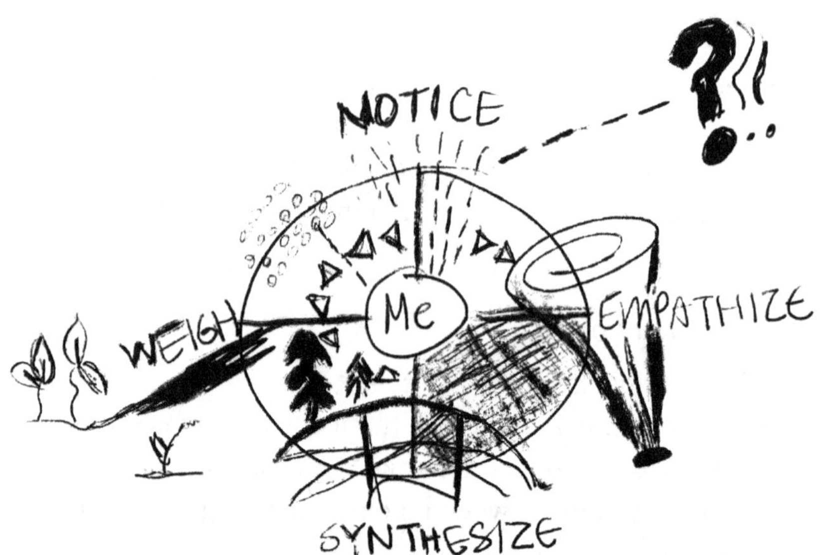
NOTICE
?!
WEIGH
Me
EMPATHIZE
SYNTHESIZE

| 4 |

Bias Coaching

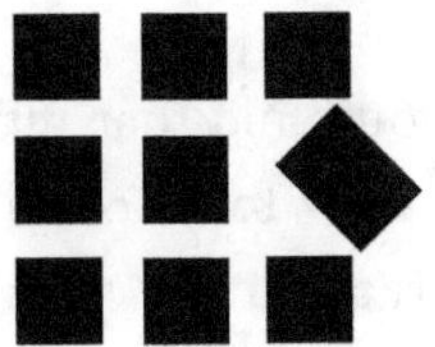

Deep listening reveals not only others, but yourself. I'm inviting you to notice what shapes your lens: your biases, assumptions, and the courage to see beyond them.

Design is never neutral. Every decision, what we include, what we omit, and how we frame the problem, carries traces of our perspectives, experiences, and unconscious beliefs. That's why the most impactful designers aren't just solving problems, they're learning to see their own lenses clearly. Bias can be a flaw to eliminate and is a constant presence to recognize, interrogate, and work with.

For designers, this means learning to coach yourself and others through blind spots. Whether you're mapping a customer journey, naming a product, or choosing default settings, your biases and those of your collaborators are already shaping the experience. The challenge isn't to become unbiased. The challenge is to become aware, and to design in a way that accounts for multiple truths, identities, and contexts.

Think of this as developing a second layer of design intuition, one that helps you pause, reframe, and get curious before jumping into solution mode. It's the shift from being the expert with the answer to becoming the guide who facilitates better questions. And just like any good coach, a designer needs to recognize when their instincts are helpful and when they might be getting in the way.

I offer tools to help you build that reflex; how to spot bias in yourself and your team, how to respond without defensiveness, how to balance pattern recognition with intentional disruption, and how to foster environments where it's safe to surface and challenge assumptions. Bias coaching, in this context, is a leadership practice, and it's one every designer can grow into.

Types of Design Bias

Bias in design isn't a moral failing. It's the natural consequence of being human. Each of us carries a lifetime of preferences, experiences, and defaults that slip into the work unless we actively hold them up to the light. The point isn't to erase bias, but to understand when it's steering the wheel without our permission. When we become aware of our internal biases, we design with sharper clarity, stronger humility, and a deeper connection to the people we serve.

Below are the biases that most commonly show up inside design teams, along with how they surface in the work and the awareness you can have in order to catch them early.

Preference Bias

You have a personal aesthetic that feels like home. Preference bias is what happens when that personal taste starts to stand in for the user's needs. We fall in love with a certain interaction, a color palette, or a micro-copy flourish, and suddenly the work begins to reflect our personality more than our audience. The risk is subtle; we create experiences that are delightful to us but disconnected from real behavior or context. Awareness means putting our taste in the backseat until we can validate that it aligns with user goals. It requires replacing "I like" with "Users need," and inviting diverse perspectives early so the design serves more than our own reflection.

Familiarity Bias

Designers are pattern-driven creatures. We rely on systems, heuristics, and past solutions to move quickly. Familiarity bias is what happens when those well-worn patterns become shortcuts that limit exploration. Instead of designing what the moment requires, we reach for the solution that worked somewhere else. The result is work that may be functional but not necessarily right or inspired. Awareness here means giving ourselves permission to diverge before we converge, even when timelines are tight. It's asking why a pattern applies, not assuming it does.

Confirmation Bias

Once we have a hypothesis about what the problem is, or worse, what the solution should be; it becomes dangerously easy to look only for evidence that supports our point of view. Designers cherry-pick research quotes, PMs share selective metrics, and suddenly it feels like everyone is nodding along. Confirmation bias narrows our field of vision. It prevents us from seeing the real story behind user behavior. Awareness means actively seeking disconfirming evidence, asking "What would make this idea wrong?" and normalizing critique that challenges, not flatters.

Solution Bias

One of the most seductive moments in design is when an idea feels so good that we skip straight past understanding the problem. Solution bias shows up when we become attached to a direction before the problem space is fully explored. It leads to ornate flows for simple issues and misalignment across teams because the design sprint outruns the problem definition. The remedy is intentional slowdown, spending more time with the tension, the context, and the trade-offs before polishing anything. This is where the compass framework (Notice, Empathize, Synthesize, Weigh) belongs; it gives you a way to hold ideas softly until they're ready to become commitments.

Feasibility Bias

Designers sometimes pre-filter ideas because we assume engineering won't build them. This feasibility bias is usually born from experience, not laziness, but it keeps us from exploring the edges of possibility. When we censor ourselves too early, we unintentionally limit the product's potential. The path out is collaboration: exploring widely, then calibrating with engineering once the landscape is clear. Feasibility should inform decisions, not dictate imagination.

Speed Bias

Product environments often move at a pace that asks designers to compress research, skip critique, and rely entirely on intuition. Speed bias makes everything feel urgent, and urgency can make even the strongest designers cut corners in ways they regret later. It leads to inconsistent decisions, usability issues, and design debt that takes far longer to fix than it would have taken to prevent. Awareness here is about choosing the smallest responsible design approach — responsible meaning ethical, inclusive, tested, and aligned with user needs. Fast work doesn't have to be shallow work.

Cultural Bias or Assumed Norms

Every designer carries their own norms around language, behavior, mental models, and even what "intuitive" means. Cultural bias creeps in when we assume our worldview is universal. It's easy to design for someone who thinks, behaves, or navigates the world like we do. It's harder to design for people whose experiences differ from ours in ways we can't see. The result can be exclusionary language, inaccessible flows, or patterns that make sense only within a narrow cultural lens. Counteracting this bias requires continual reminders to ask: "Whose experience am I missing?". Testing with diverse users and collaborating with accessibility and DEI partners ensures the product includes not just the majority, but the margins.

Sunk Cost Bias

Once we've put hours into a direction, it's painful to consider changing course. That's sunk cost bias, the tendency to stay committed to the wrong idea because abandoning it feels like failure. Teams cling to features because they required effort, not because they provide value. Designers polish concepts that should have been gently retired. Awareness means being willing to let go. Prototypes should be cheap and disposable. Pivoting should be normal, not shameful.

Expertise Bias

With experience comes intuition, and an overconfidence that can block growth. Expertise bias leads designers to trust their instincts over data, to dismiss junior voices, or to resist new methods and tools because the old ones "worked fine." This bias is especially common among senior designers who have been rewarded for their taste. Awareness means staying teachable, inviting newer voices into critique spaces, and continually refreshing your perspective. Expertise should be a contribution, not a constraint.

Optimism Bias

Optimism is a beautiful trait in designers; we imagine better futures. But optimism bias can lead us to overestimate user understanding, assume friction will magically resolve itself, or design ideal-state flows instead of real-world ones. I've witnessed that we think users will "figure it out." They don't. Awareness means validating comprehension, not just usability, and remembering that real users bring interruptions, stress, disabilities, and imperfect attention.

Bias Grounding

Bias is not the enemy in design. It's the starting point. Every designer comes into the work with instincts shaped by experience, taste, culture, upbringing, and past projects. Pretending we can design without bias is unrealistic; acknowledging it is where our best work begins. When we see bias not as a flaw but as a signal, we give ourselves permission to pause, question our assumptions, and widen the lens beyond our own perspective.

The tricky part is that most bias lives under the surface. It shows up in the quick decisions, the patterns we reach for without thinking, and the internal dialogue that nudges us toward what feels comfortable. I often think of it as two voices, one of each of your shoulders. One says: "You know this. Design for people like you. It's faster, familiar, and probably fine." The other whispers: "Hold on. Who else might need this? What don't you know yet?" The tension between the two is where thoughtful design lives. We're not trying to silence either voice; we're trying to stay aware of which one is steering us.

Bias isn't always harmful. In fact, some of our most reliable instincts come from years of pattern recognition. Familiar heuristics help us move quickly when time is tight. Experience helps us lean toward safer, clearer solutions in high-stakes environments. Even our aesthetic preferences can spark creativity within constraints. The goal isn't to purge ourselves of every bias but to recognize when bias is guiding us well and when it's limiting our view. Awareness is the differentiator.

Which is why intentional reflection must be part of your practice. Bias becomes problematic when it operates unnoticed, shaping decisions that unintentionally exclude, oversimplify, or reinforce our own worldview. When we stop and examine the stories, we're telling ourselves, we make space for more inclusive choices. We create better products, but we also grow as designers.

Bias-Busting Questions

Here are powerful prompts designers can use to surface internal bias and reconnect to intention. These questions are not meant to shame, but to expand awareness and offer a reset point when instinct takes over:

1. **Who am I designing for, and who might I be leaving out?**

   Consider age, ability, culture, tech access, language, and lifestyle. What assumptions am I unconsciously layering in?

2. **How does this design reflect my own preferences?**

   Am I choosing features, visuals, or language that resonate mostly with me? What alternatives could better serve a broader range of users?

3. **What do I mean when I say something is "intuitive"?**

   Intuitive to whom? Do I have research to support that assumption?

4. **Am I stereotyping or oversimplifying personas?**

   Are my personas representing real variation, or are they flattening an entire group into one narrative?

5. **Whose perspectives are missing from my process?**
   Whose feedback would challenge my assumptions in a
   healthy way?
6. **Am I defaulting to familiar patterns because they're
   easy?**
   Are their additional patterns or solutions worth considering
   before I converge?
7. **Could my testing choices be reinforcing my own
   worldview?**
   Are my research participants too similar to me or my team?
8. **How might different contexts or constraints change
   this design?**
   What happens on low bandwidth, old devices, or with different physical abilities?
9. **How am I defining success, and is that definition inclusive?** Are my metrics centered only on typical users, or do
   they reflect diverse needs and outcomes?
10. **What would I change if I were designing for someone
    completely unlike myself?**
    How would I shift language, hierarchy, interaction, or tone
    if my mental model were intentionally flipped?

These questions aren't meant to slow designers down, they're
meant to make us more intentional, more inclusive, and more
confident in the decisions we make. Awareness doesn't replace instinct; it strengthens it.

Bias Coaching with Soul

Bias coaching isn't a one-time event, it's a loop. A continuous
awareness practice that flows between the user, the team, and the
self. As designers, we sit at the intersection of perspectives, which

means we're uniquely positioned to spot when assumptions are guiding choices more than evidence or intention. This loop starts with the user, noticing whose needs are centered and whose are missing. Then it moves through the team: surfacing shared blind spots, language habits, or default behaviors that may exclude or oversimplify. Finally, it comes back to the self, the hardest and most important place to look. Designers can lead through example; pausing to ask clarifying questions, naming when bias may be influencing direction, and modeling curiosity rather than certainty. Bias shows up in design processes, research framing, naming, prioritization, and even in how we define success. Coaching around bias doesn't mean being right, it means helping the team stay in a loop of reflection, inclusion, and course correction.

Inclusive Design Principles

Designing for diverse users is a cornerstone of creating impactful, accessible, and user-friendly experiences. As the world becomes increasingly connected, ensuring that your designs meet the needs of people from different backgrounds, abilities, and experiences is not just ethical, but essential to the success of your product. Designing inclusively means recognizing and embracing the diversity of your user base, which includes various physical abilities, cultural contexts, and technological access.

The first step in inclusive design is to genuinely understand the diversity of your users. This means considering not just the typical user, but also those with disabilities, those from different cultural backgrounds, and those with varied levels of technological literacy. Employ user research methods such as interviews, surveys, and ethnographic studies to gain deep insights into the needs of diverse user groups.

Make sure that your designs are usable by people with a range of physical abilities, such as those with limited mobility or color blindness, by offering keyboard navigation, high contrast themes, and text resizing options. By adhering to universal design principles, you're not just making your product more accessible to a broader audience but also enhancing usability for everyone. Inclusive design doesn't just mean adding accessibility features; it also involves considering usability in the broadest sense.

I've witnesses edge cases leading to innovation in the realm of accessibility. Often, the features or accommodations designed for users in edge cases spark innovations that ultimately benefit everyone. For example, voice-enabled interfaces were initially developed to support users with mobility impairments, but today, they're widely used for a variety of tasks by all types of users. Similarly, designing with a focus on low-bandwidth internet access or devices with minimal processing power creates products that work efficiently for everyone, even in high-traffic or resource-limited environments.

Incorporating inclusivity and accessibility into the design process isn't just a nice-to-have; it's a necessity in a world that increasingly values equity, diversity, and user-centricity. When we go beyond the mainstream and focus on marginalized groups, we create products that not only serve a wider range of users but also push the boundaries of innovation. By challenging the status quo and thinking inclusively, we can create experiences that feel personal, empathetic, and impactful to all users, ultimately building better, more resilient products.

Build a Bias-Resistant Design Process

Biases, whether implicit or explicit, can shape design decisions in subtle ways that may unintentionally exclude or marginalize users. To create products that are fair, accessible, and effective, designers must actively work to identify and reduce biases throughout the design process. This requires adopting strategies that promote collaboration, transparency, and continuous learning. Below are some key practices for building a bias-resistant design process:

1. Get feedback from users!
2. Be open; talk about it.
3. Collaborate, ask people who are not like you what their thoughts are.
4. Get feedback from others in the organization.
5. Get feedback from someone who is not a current user but could be someday.
6. Document your decisions!

As you uncover biases and shape the design, document your rationale. Record why certain design choices were made, who was involved in those decisions, and the data or assumptions that supported them helps create transparency. Over time, this documentation serves as a valuable tool for reflecting on past decisions, identifying patterns of bias, and improving future design processes. This is a great central tool for bigger team decisions but also is a great to document within specific design initiatives. This will go a long way to create a culture of bias awareness which help encourage these important questions for us to address with each other.

Bias has truly been around for as long as humans have existed, shaping decisions and perceptions. Have you ever found yourself designing a feature because it just "feels right"? Or making a choice because it aligns with how you or your close circle would naturally use the product? The funny thing is that bias usually slips in under the guise of good intentions. For example, you might be trying to make the interface super easy to use. But what does "easy" mean? And to whom? Without meaning to, you could be designing an experience that fits one particular user type while unintentionally excluding others.

Product design has this wonderful paradox; it's about solving problems for real people, but we often approach it within our own little bubbles of perspective. Breaking out of that bubble is the first step toward tackling bias. But here's the catch: we don't always see the bubble because it feels like home. Our assumptions, habits, and even our favorite design patterns all live inside this bubble, shaping our choices and coloring our decisions. Be aware of those subtle whispers of bias, the ones that seem harmless at first but can grow into barriers for many users. As you do this, you'll be equipped with strategies to design in a way that isn't just for people like you, but for a true diversity of users.

How I've Helped My Teams Navigate and Reduce Bias

An important design ritual are design critiques or also known as "crits." I believe that critique should center the problem, not the person. So instead of asking designers to defend their work, I ask the team to explore the intent behind it: What problem are we solving? Who are we designing for? Whose experience might be missing? This structure removes ego from the conversation and gently exposes the assumptions we're making. It creates room for designers to name their own blind spots without fear, and it reinforces that reducing bias is a shared responsibility, not an individual flaw.

Research has also played a foundational role in how I help teams counteract their biases. I've led teams in environments where the "default" user was often assumed to be someone just like us, which, of course, is rarely true. I pushed us to look beyond obvious or convenient user groups. I reframed research not as a burdensome step, but as a practice of listening deeply and honoring the diversity of lived experience. This work often dissolved confirmation bias before it had a chance to calcify.

Surface biases by simply widening who is in the room. Invite voices from customer support, sales, success, accessibility, product management, engineering, and real users into the design process. These diverse perspectives bring nuance that no amount of theoretical persona work can replicate. They reveal where our defaults are too narrow or too assumptive and push the team to consider the full breadth of human need.

As a leader, I've tried to model the distinction between intent and impact. Teams often get attached to the intention behind their work, what they meant to do, who they thought they were helping, and lose sight of how people actually experience the product. I regularly ask the team to look at the gaps between intention and outcome without judgment. This shift gave designers permission to acknowledge when their instincts were off and helped them approach feedback with openness rather than defensiveness.

Accessibility is a pillar of my leadership philosophy. I've never waited for legal pressure or PM direction to prioritize accessibility; I've framed it as an act of empathy and inclusion. I have taught teams to run accessibility checks, use screen readers, test color contrast, and consider situational disabilities. This prevented exclusionary design choices from becoming embedded in the foundation of our products.

Finally, I strive to lead with vulnerability. I tell my teams when I'm relying on instinct. I openly ask, "What am I missing?" I invite them to disagree with me, and I genuinely listen. This kind of humility creates psychological safety and gives designers permission to challenge their own assumptions as well as each other's. When leaders show their own bias, it becomes easier for everyone else to acknowledge theirs.

Across all these approaches, the common thread is awareness — creating practices, spaces, and cultures where bias can be seen rather than hidden. Not eliminated but understood. Not judged but examined. And in that awareness, design becomes more human, more inclusive, and far more powerful.

Carry This Forward

Bias isn't the enemy, it's information. When you pause long enough to examine your instincts, you create space for more inclusive, intentional design. Carry forward this awareness as a practice of curiosity, not perfection.

Chapter Reflection

Reflection Prompts

- What assumption did I make today?
- What surprised me in user feedback?
- Who challenged my thinking this week?
- Where have your biases shaped design choices without you realizing it?
- When have you paused a team to challenge assumptions? What was the outcome?
- What helps you notice bias in yourself with compassion, not shame?

Artistic Expression

Create a split composition; one side represents a lens of bias, the other, a widened view. Use color, blur, or imagery to show distortion vs clarity.

Soulprint Words

Choose 3–5 words that capture what this chapter unlocked in you.

Examples: Aware, Disruptor, Conscious, Curious, Brave

Soulprint Thread

Bias understanding reveals your capacity to *design with awareness and courage.*

Carry forward a reminder: *"I hold space for multiple truths."*

Cultivate Curiosity

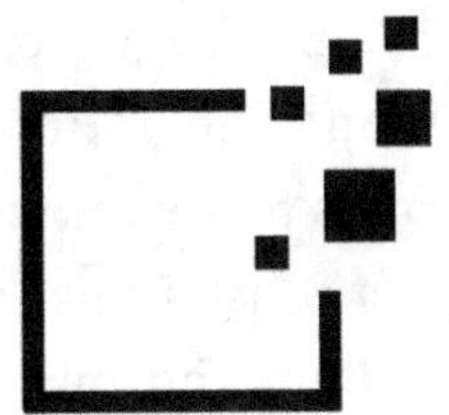

Awareness opens the door. Curiosity walks through it. Staying curious keeps you adaptable, inventive, and fully alive in your work.

You've heard the term, "cat herding," right? I use this term to describe trying to organize a group of highly independent, unpredictable, and often uncooperative individuals, kind of like a bunch of cats each doing their own thing. Sound familiar? If you've ever worked on building a product, you know exactly what this feels like.

Cat herding is the ultimate exercise in organized chaos! It's about doing the impossible, bringing order to a situation that seems determined to stay delightfully chaotic. And in the design world, this perfectly describes those moments when you're tasked with getting cross-functional teams: engineering, marketing, product management, you name it, to align on a shared vision.

Each team, like each cat, has its own personality, its own set of priorities, and sometimes even its own language. Engineering might be laser-focused on feasibility, product management is obsessed with timelines, and marketing is dreaming about how the product shows up to the world. Everyone is moving in different directions, and as a product designer, you're tasked with understanding those perspectives while maintaining your creative grace and keeping the bigger picture in mind.

The tricky part? There's no magic song that gets all the cats moving in perfect formation. There's no single solution for getting cross-functional partners to see things the same way or agree on what happens next. It's a constant back-and-forth, a balancing act between listening, guiding, and making sure everyone feels heard and valued.

Here's the good news — while you can't control the cats, you *can* control how you interact with them. And that starts with understanding yourself, your role in the chaos, your triggers, your strengths, and the way you respond under pressure. By knowing yourself, you can shift from reacting to gracefully navigating. Instead of forcing the cats into one tidy line, you become the person who guides them toward a shared direction, even if they take a few detours along the way.

The mindset required to navigate the cat herd is rooted in curiosity, empathy, and flexibility. Curiosity helps you understand each "cats" needs and perspective; empathy keeps you connected to the humans behind the chaos, and flexibility allows you to pivot without getting overwhelmed.

Get Curious!

There's an old saying, "curiosity killed the cat," but it didn't start that way. Earlier versions warned that *care* (meaning worry) could be just as dangerous. Over time, curiosity took its place, and later, people added the reminder that satisfaction brought the cat back. In design, that evolution matters. Curiosity isn't the risk. Staying stuck in worry is. Curiosity isn't dangerous, it's the only path to understanding, innovation, and ultimately, satisfaction.

In the world of design, curiosity is the lifeblood of exploration and invention. It pushes you beyond the surface, urging you to dig deeper, challenge assumptions, and constantly ask, "what if?" This requires presence, energy, and a willingness to keep your mind open even when the path feels unclear.

Curiosity empowers designers to question the status quo. Instead of accepting things as they are, a curious mind probes and pokes at conventions, asking why something works the way it does and whether it should. By challenging long-held assumptions, designers uncover opportunities for improvement that others might miss. Never assuming anything keeps your mind, and the minds of those around you, open to possibility. It helps break the, "we've always done it this way" cycle and creates space for new approaches to emerge. Design, after all, is a field of constant evolution, and curiosity is what keeps you moving rather than stagnating.

Curiosity also leads designers into new territory. It encourages exploration, experimentation, and play. A curious designer might test unfamiliar tools, borrow ideas from unexpected industries, or incorporate learnings from nature, architecture, or AI. This exploratory mindset often reveals surprising connections and creative breakthroughs. A gaming interface might inspire a financial app, or traffic patterns might spark ideas for streamlining an e-commerce flow.

Staying curious cultivates openness. It helps designers embrace the unexpected; the user comment that shifts the direction of a project, the small insight that grows into a major product improvement, or the experiment that leads somewhere better than the original plan. Curiosity keeps the design process dynamic rooted in continuous learning, iteration, and discovery.

Instead of executing tasks on autopilot, curious designers stay in relationship with the work. They adapt. They inquire. They adjust. And in a landscape where user needs, technology, and market expectations are always moving, curiosity isn't just helpful — it's essential.

Understand Yourself

To create great work, designers must first understand their own strengths, weaknesses, biases, and tendencies. This includes discussions on emotional intelligence and self-reflection. Understand how introverts and extroverts approach design challenges differently. How does understanding your own personality help you shape your working habits or collaborate more effectively with others? There is no playbook to this, beyond pure mindfulness with a growth mindset. Be authentically you! How do your own values influence what you're curious about and your design decisions.

Curiosity in Practice

Schedule time specifically for exploration and creative play. Whether it's a few hours a week for side projects, learning new tools, or following a random design interest, having "curiosity time" built into your routine helps keep you in a state of creative flow. During this time, let yourself follow your curiosity without worrying about deliverables or deadlines.

In practice, I've seen many designers shy away from curiosity out of fear of making mistakes. I'd challenge you to be curious about why you think that. As humans, we have been trained and rewarded for being a certain way and we need to unlearn these behaviors. Explore how embracing curiosity means welcoming failure as a learning opportunity, not a setback.

Ultimately, curiosity is the spark that leads to innovation. It encourages designers to not only ask "why" but also "why not?" This fearless pursuit of new ideas and better solutions is what drives progress. Whether it's reimagining an entire user interface or making a seemingly minor design change that transforms the user experience, curiosity propels designers to push boundaries and find creative, unexpected answers to complex problems.

In a world where technology, user expectations, and business goals are constantly shifting, curiosity is the trait that keeps designers ahead of the curve, helping them uncover the next great idea before anyone else sees it. It's the secret ingredient that turns good design into exceptional design — always challenging, always exploring, always innovating.

If curiosity is the mindset, here's a simple order of operations for putting it into practice in your design work:

1. **Start by understanding yourself.**
   Notice your defaults, triggers, design preferences, and instincts. Ask: Where am I making assumptions? What do I tend to overlook? Self-awareness is the foundation of unbiased curiosity.
2. **Slow down long enough to question your first thought.**
   Don't accept your initial idea or interpretation as the final word. Pause and ask: What else could be true? What am I not seeing yet?
3. **Dig deeper into the problem.**
   Go beyond surface symptoms. Probe the "why behind the

why." Curiosity thrives when you linger in the problem space instead of rushing to solutions.

4. **Challenge assumptions and the status quo.**
   Flip your thinking. Ask: Why do we do it this way? Should we? What would happen if we tried the opposite? This breaks the "we've always done it this way" cycle.

5. **Explore beyond your comfort zone.**
   Experiment with new tools, methodologies, or inspirations. Look outside your domain — nature, architecture, gaming, AI. Curiosity grows in unfamiliar territory.

6. **Stay open to unexpected insights.**
   Let new information change your direction. When user feedback surprises you or a teammate offers a perspective, you didn't anticipate, treat it as fuel, not friction.

7. **Iterate with intention.**
   Stay in conversation with the work. Adjust, test, and refine. Curiosity isn't linear — it loops, evolves, and deepens with each iteration.

8. **Reconnect to the humans behind the product.**
   Ground your design decisions in empathy. Ask: What does this feel like for the person using it? What might they need that I haven't considered?

Connect Perspectives

Let's say you are a designer building a website that has six department heads all of whom have equal priority and power and you've been given the task to listen to requirements to make the best experience for users. These stakeholders may or may not have a keen understanding of the user, so first, you need to get curious about that and find data to help understand the current situation (if there is a current website you are rebuilding). For instance, I've seen things like carousels being used to not answer the hierarchy question and instead lump all of things in a bucket to serve organizational hierarchy, which is not a user centered approach. Instead, listen to every stakeholder and what they want, ask a lot of questions to what they are looking to achieve. Then, learn what a great experience would look like from a user point of view. The solution will be nothing that anyone could envision, this will truly be an innovation and will bring your unique power of solutioning to complex problems. In a sense, this is the cat herding that can turn cats into synchronized swimming champions. Curiosity is the thread that ties cross-functional partners together.

Designers as Translators

When engineering, product, and design feel like they're moving in different directions, it's often not because of misalignment on the goal, it's because of how each team gets there. It's like we are all our own breed of cats!

Designers can bridge these dialects by listening closely and framing ideas in terms that resonate across disciplines. When engineers push for speed, explain how a cleaner interface can help users move faster too. When product wants to ship quickly, highlight which elements will drive engagement now, without sacrificing long-term vision. The key is reframing — not compromising — your intent so others can see its value in their terms.

Here are practical ways to translate across teams:

1. **Find the Shared North Star**

    Remind the team of the common mission — better user outcomes, growth, or reliability. Anchor conversations in that goal so everyone sees their work as contributing to the same success.

2. **Speak Engineer**

   Frame new features in terms of reduced errors, improved performance, or fewer server calls. When you show how a design idea optimizes something, you're more likely to earn buy-in.

3. **Paint the Business Picture**

   Product teams need to see the return on investment. Share how your design direction opens new revenue or market potential. Help them see your vision through a business lens.

4. **Use "Yes, and…"**

   When engineering says something's a heavy lift, respond with agreement and an addition: "Yes, and here's how we might simplify it." Respecting constraints opens space for collaboration, not conflict.

5. **Make Tradeoffs Visible**

   Decisions get easier when the impact is clear. If a richer experience increases load time, lay that out transparently. Let the team weigh options with shared understanding. System constraints and speed-to-market pressures are real. Treat them as design constraints. Ask, "How might we create a great experience within these boundaries?" That shift builds trust.

Curiosity and Collaboration

Curiosity doesn't just help a designer work better alone; it also enriches collaboration. A curious designer asks questions not just of their users, but of their teammates as well. They engage with engineers, marketers, and product managers, seeking to understand different perspectives and how they can enhance the design process. This curiosity about others' expertise often leads to more cohesive, cross-functional collaboration, as the designer becomes the bridge between different teams, driving alignment and innovation.

Curiosity also requires nurturing. Working 80 hours on the same problem week after week is going to cloud your brain. You will need to expose yourself to perspectives outside your usual bubble, in your work and personal life both. Talk to people in different fields, learn about their processes, and ask how they approach problem-solving. Insights from other disciplines can lead to new ways of thinking and help you see your own work in a fresh light.

Curiosity thrives in collaborative environments. Engage with colleagues who think differently than you, brainstorm together, and listen to their ideas and feedback. Being open to collaboration helps you learn from others and exposes you to new ways of thinking. Group brainstorming sessions can often lead to surprising, curiosity-driven ideas that you might not have arrived at alone. Stay curious about how users interact with your designs. Spend time analyzing user feedback, observing behavior through usability testing, and asking users thoughtful questions about their experience. When you approach user feedback with curiosity, you can uncover underlying problems or unexpected insights that lead to better design decisions.

Building Bridges with Product Managers

Product managers orchestrate priorities across business goals, user needs, and technical feasibility. Partnering well with them means understanding their world. Early collaboration creates joint ownership and reduces last-minute friction. Learn what's driving product decisions. Competitor moves, user research, or market demands may shape feature direction. When you align design strategy with product context, your work lands stronger — especially when product is invited into ideation before designs are polished.

Embracing Empathy for Engineers

Engineering teams are balancing limited resources, technical debt, and future-proofing. The more you understand their challenges, the better your design will serve everyone. Pick up the basics of how things are built. Knowing a little engineering vocabulary helps you hold more productive conversations and build trust. Invite engineers into brainstorming and early review. Their input often leads to more creative, efficient solutions. It also prevents rework later.

Constraints Are Invitations to be Curious

When design challenges arise, it's easy to see constraints as frustrating roadblocks. But what if you saw them as invitations instead? Constraints offer structure, focus, and friction — fertile ground for curiosity to grow. Start working with them instead of against them.

As a designer, tension is part of the work. But tension doesn't mean you're stuck, it means you're growing. When you lean into creative friction, you build resilience. Every time you face a constraint and choose curiosity, you strengthen your creative muscle.

Practice it. Join a challenge that pushes you. Collaborate with people who think differently. Take on the project that makes you nervous. Each moment of discomfort is a chance to grow your capacity for inventive, thoughtful design.

The next time you feel resistance in your process, pause. Get curious. Ask, "What is this constraint trying to teach me?" Because the designers who thrive are not just the most talented, they're the ones who stay curious, especially when things get hard.

Reframe your Perspective

Rather than treating constraints as barriers, treat them as prompts for creativity. What can you invent within the boundaries you've been given? A tight budget might lead you to embrace a minimalist approach that ends up being more elegant and effective. Constraints can sharpen your thinking and invite surprising solutions.

Use Boundaries to Spark Ideas

Boundaries can bring clarity. Try setting self-imposed limits, three colors, one typeface, or a specific user constraint, and use those to drive focused exploration. Much like a poet writing within the rules of a haiku, constraints can guide you to powerful ideas you might not have uncovered otherwise.

Carry This Forward

Curiosity turns chaos into discovery. It's what transforms constraint into creativity and difference into dialogue. Carry forward your willingness to ask, "what if?" … especially when the answer isn't clear.

Chapter Reflection

Reflection Prompts

- What ignites your curiosity lately?
- Where do you fall into "we've always done it this way" thinking — and how might you reimagine it?
- What does a beginner-mindset look like in your process?

Artistic Expression

Draw a curiosity trail: loops, questions, detours, wonderings. Let it be nonlinear. Let it surprise you. Show where your energy wants to go.

Soulprint Words

Choose 3–5 words that capture what this chapter unlocked in you.

Examples: *Playful, Explorer, Adaptive, Open, Experimental*

Soulprint Thread

Curiosity sources reveal your creative energy.

Carry forward one wild idea, sketch, or phrase that sparks wonder.

"Curiosity is the catalyst for transformation."

| 6 |

From Insight to Expression

You explored how curiosity helps you listen more deeply, ask better questions, and see more of the world. Now, let's learn how conviction turns all that insight into something real and shareable.

Diagrams, user journeys, and prototypes communicate faster than slides or specs. A well-illustrated concept helps everyone see the problem and their role in solving it. This is the moment when all that listening and intention begins to take shape in the form of an experience, a design.

Now it is time to put all the learning and intention into producing the experience. Because you have done the work to understand the perspectives, the user, and the business, your skillset becomes the tool that helps a team, a client, or a business move forward.

Design in any flavor, from low fidelity to high fidelity mockups to prototypes, lets you test ideas quickly, show how different priorities can coexist, and even gather feedback without broader investment. If the product team wants quick results, show them how a prototype can meet that need while still allowing room for innovation. Engineers appreciate seeing a concept before investing in code, and product managers value an early glimpse of how users might experience the solution. Early visuals create alignment, narrow ambiguity, and give the team something concrete to react to.

But here is the part people often overlook — before any of that happens, before the alignment, the feedback, the collaboration, there is a moment every designer faces alone. To bring an idea into the world, you must first go inward. You must sift through the noise, explore possibilities, and follow threads no one else sees yet. This is where the designer enters the quiet, creative isolation that shapes the early soul of the work.

Navigating Design Decisions with Empathy

Before you dive into design decisions, take the time to create user journey maps. These visual representations of user interactions with your product illuminate their needs at every touchpoint. By understanding the user's journey, you can identify friction points that need addressing and highlight areas where your shiny new feature might just be overkill.

Establishing regular feedback loops with actual users can feel like having a cheat sheet for your design decisions. Their reactions will guide you toward empathy-driven compromises. If users' express frustration over a clunky feature, it is time to either rework it or consider letting it go altogether. Remember, design is not about making something pretty; it is about making something that serves.

Embrace a culture of testing and iteration. When tensions rise during the design process, do not just bulldoze ahead with your original vision. Test your ideas with users to see what resonates and what falls flat. Each iteration is an opportunity to enhance empathy. By learning what works for users, you are continually honing your design compass.

Isolation, Ambiguity, and Defining the Box

The Isolation

There is something magical, and mildly terrifying, about working alone as a designer. It is like stepping onto a stage with a packed audience, armed with only your thoughts and a blank canvas. No partner to lead, no outside voice to offer feedback. Just you, your ideas, and the infinite possibilities ahead.

Isolation in design can also be a beautiful thing. It allows you to dive deep into the unfiltered recesses of your mind where creativity is not yet polished by outside input or bogged down by constraints. In these moments, you get to play. Really play. It is just you and your thoughts.

But let's be real. There is a certain kind of madness that comes with being alone in your creative headspace for too long. You start to have conversations with yourself, and, quite frankly, your brain can become your own worst critic and best cheerleader all at once. One minute you are convinced you have just had a stroke of genius, and the next you are looking at that same idea wondering if it is brilliance or just the result of too much coffee.

Isolation, at its best, is where the deep stuff happens. It is where you dig into the raw material of creativity. There are no distractions, no noise, just you peeling back layers of an idea until you hit something profound. Or at least something that looks profound at 2 a.m.

In this space, introspection becomes your best friend. You start asking the tough questions: What am I really trying to say with this design? Does this interaction evoke the feeling the end user would want? Does this color evoke the right emotion? Why did I choose this element? Does this hierarchy of elements reflect my biases? Did I just design something pretentious?

But the deeper you go, the more vulnerable you become to the rabbit hole of overthinking. This is where isolation turns into both your greatest ally and potential nemesis. You can easily get lost, stuck refining an idea to the point where you are not even sure what you are refining anymore. Isolation has this funny way of making you question everything. Does this design even matter? Am I really solving a problem, or am I just indulging my own aesthetic whims? Am I even a real designer?

These internal dialogues are part of the process. It is here, in isolation, that you get to interrogate your instincts, chase wild ideas, and explore concepts without the fear of immediate judgment. It is the creative equivalent of singing in the shower. No one is around to hear you miss the high notes, so why not belt it out? The shower is where brilliant ideas come for me. Also, sleep. I can ask my brain questions, and I wake up with answers. Knowing your own natural creative rhythm and source can help you fuel solutions.

In isolation, you can experiment without judgment. It is where bold ideas are born, the kind that make you think you cannot believe you are about to try this, but here we go. Sometimes it works, sometimes it does not, but that is not the point. The point is that solo work gives you the space to push boundaries, take risks, and explore ideas you might not dare to share in a collaborative setting right away. This is where the soul of a designer really starts to come out.

The thing about isolation is that it is messy. It is not a neatly packaged process where every idea makes sense or has a clear path to execution. It is raw, and it can be uncomfortable. But that is where innovation lives, in those moments where you are forced to sit with an idea long enough for it to become something new, something unexpected. We call this "Discovery".

Eventually, you must step back into the world of collaboration, where those bold ideas get refined and shaped by feedback. Isolation gives you the raw material, the bold strokes, the wild ideas. Collaboration adds the polish.

Ambiguity to Definition

Ambiguity is often seen as the enemy in design, especially when deadlines, budgets, and expectations loom large. But in my experience, ambiguity holds immense potential. It is the realm of possibilities, where ideas roam freely, unbound by strict rules. Ambiguity invites us to explore, experiment, and redefine the norms. Personally, I thrive on it. The more ambiguous the problem, the more it excites me. But I know this is not true for everyone. For some, ambiguity feels overwhelming and paralyzing.

As designers, I believe we must learn to harness this ambiguity, treating it as a canvas rather than a void. Ambiguity is not about a lack of direction; it is about the freedom to carve a new path. But to do this effectively, we must balance it with definition.

For me, ambiguity fuels my creative energy. When a project starts with little definition, it is an opportunity to envision what could be, uninfluenced by boundaries. But this is not everyone's approach. Some designers crave more structure from the start. Recognizing your natural orientation toward ambiguity is important. Embrace it if it excites you. Seek more structure if that is what you need.

By seeking that structure, I do not mean relying on others to make design decisions. Here is where the scope of design is ambiguous by itself. Earlier in the materials, I mentioned that Product's job is to set a success metric and let the design team go. When you understand your own response to ambiguity, you can make it a tool rather than a hurdle.

Defining the Box

Ambiguity alone will not get you across the finish line. At a certain point, definition must step in. The more clearly a project is defined, the more likely it is to stay focused and aligned with its purpose. While it may sound counterintuitive, freedom in design often thrives within boundaries. You cannot explore meaningfully unless you know where the edges are.

Defining the box is not about limiting creativity; it is about giving it something to push against. A well-defined box creates momentum, reduces noise, and helps teams move from possibility to progress. So, what does it actually mean to define the box?

Scope and Constraints

Defining the box means making the scope visible. It includes what you are solving for, what you are not solving for, and the conditions that shape the work. This might include features, timelines, budgets, technical limitations, legal or accessibility requirements, and the level of fidelity or risk the project can support.

Clear constraints help everyone understand where to focus their energy. They reduce false starts and prevent the slow drift into scope creep or endless exploration. When the box is unclear, teams often spend time debating direction instead of building momentum.

Just as importantly, defining the box is a shared act of alignment. It gives designers, partners, and stakeholders a common understanding of what success looks like right now. Not forever. Not in some ideal future state. But in the context of this moment, this release, this decision.

The box is also not static. As you learn more, through research, testing, or real-world use, the edges may shift. Defining the box is not a one-time declaration. It is an ongoing practice of revisiting assumptions, adjusting constraints, and deciding, together, when it is time to expand or redraw the boundaries.

When the box is clear, you can spend less time guessing and more time creating. The work becomes more intentional, more grounded, and ultimately more impactful.

Expression Is a Translation Skill

Expression is not just output. It is translation. You are translating research into form, strategy into experience, and ambiguity into something a team can react to. When your work is early, it is also fragile. It is easy for other people to project onto it. They see a rough wireframe and assume it is a final opinion. They see a prototype and start debating visual design. They see a journey map and forget the human underneath it. That is why expression is not only craft but also framing.

Before you share anything, ask yourself: "What question is this artifact answering?" A journey map answers, "Where does this experience break down for someone?" A wireframe answers, "What is the structure and sequence?" A prototype answers, "What does it feel like and where does it fail?" When you label the purpose, you protect the work from being misread.

A simple way to do this is to attach a short caption to every artifact you share, even internally:

- **What it is:** journey, wireframe, prototype, concept model
- **What it is not:** final UI, approved solution, complete flow
- **What feedback you want:** comprehension, risk, usability, feasibility
- **What decision it supports:** what we build next, what we test, what we cut

This small act changes the room. It keeps you from getting dragged into premature debates, and it keeps the team anchored in the right layer of conversation. It also teaches your partners how to work with design. Many people have opinions, but fewer people know how to give useful feedback. Your framing helps them.

This is also where conviction shows up. Conviction does not mean stubbornness. It means you can say, "Here is what I am confident in, here is what I am exploring, and here is what I still need to learn." Designers who can name that difference build trust quickly. They do not pretend to know everything. They do not apologize for iterating. They invite the team into a process that has structure.

If you want a quick mantra for this phase, use: **show, name, invite.**

Show the work. Name what it is and what it is not. Invite feedback that advances it.

Carry This Forward

Conviction gives your ideas backbone. It is how curiosity takes shape and moves the world forward. Carry forward the courage to create with clarity, to sit with ambiguity and isolation long enough to find something true, and then to bring that truth into collaboration with others.

Chapter Reflection

Reflection Prompts

- What are you currently holding back from saying or showing in your work?
- What's one idea or point of view you've been holding back, waiting for permission to share?
- What would it look like to design like your voice matters?
- Where does isolation, ambiguity, or "defining the box" show up in your current project, and how could you move through them with more conviction?

Artistic Expression

Illustrate or collage a flame. Let it be bold, textured, radiant. This is your design fire. Add to it the names of the ideas you want to protect and the ones you're ready to let burn. If you want, sketch a box around the flame and draw where the edges of constraint actually help it burn brighter instead of going out.

Soulprint Words

Choose 3–5 words that capture what this unlocked in you.
Examples: *Courageous, Challenger, Truth-teller, Unshakable, Honest*

Soulprint Thread

We have revealed your fire. Carry forward a mark, phrase, or image that captures your creative fire – "I design with clarity, even when it's uncomfortable." In your Soulprint, this becomes the spark that shows what you stand for when you bring ideas out of your head and into the world.

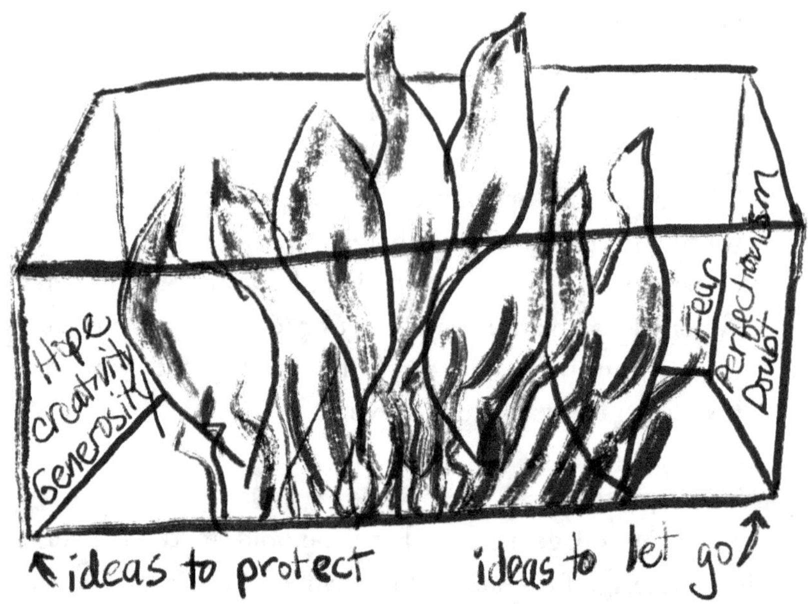

Hope
creativity
Generosity
Fear
Perfectionism
Doubt
ideas to protect
ideas to let go

| 7 |

Art of the Edit

Y ou've learned how to listen deeply and bring ideas into form. Now comes a quieter, more demanding skill: deciding what stays. The art of the edit is where design matures, where clarity is earned not by adding more, but by choosing with intention what deserves to remain.

Editing is not just trimming. It is judgment. It is knowing the difference between what is interesting and what is necessary. It is the ability to hold the user's needs in one hand and the business reality in the other, then decide what earns a place in the final experience.

Simplicity vs. Complexity in Design

Every designer has felt it, that urge to keep adding, tweaking, embellishing. After all, more must mean better, right? Well, not always. In the world of design, simplicity can carry as much weight as complexity. Knowing when to strip things down or dial them up is the art of restraint. It is the difference between elegance and excess, clarity and clutter, usability and confusion.

Design is not a buffet where more is inherently better. It is a dish where every ingredient counts, and each addition should serve a purpose.

Restraint is not just about what you leave out. It is about making sure what you put in has room to matter. Restraint turns a feature, an interaction, or a visual element into a powerful moment instead of one in a crowd. When you let simplicity take the stage, it can make a design feel honest, effortless, and, most importantly, effective. It is not about deprivation, it is about selection, ensuring that each element has room to breathe and resonate with users.

Clarity Over Clutter

One of simplicity's greatest virtues is clarity. When there is less competition for a user's attention, the essential parts of an experience can stand out, guiding users intuitively toward what they need. In interfaces especially, simplicity is a beacon of ease, making it clear what to do, where to go, and how to navigate seamlessly.

Simplicity Feels Honest

Have you ever seen a design that feels overly dressed up, as if it is trying a bit too hard to be cool? Often, that comes from unnecessary complexity, which can feel inauthentic. Simplicity has an inherent honesty to it. It is less about showing off and more about creating a smooth, straightforward experience.

Building Trust with Users

Complexity can overwhelm or even alienate users. When designers pare things down to the essentials, it builds trust, giving users confidence that they know exactly what to expect and how to get it. Trust comes from reliability, and simple, intuitive design is often the fastest path to that comfort zone.

Timelessness

Ever notice how the simplest designs often seem to age the best? Trends come and go, but simplicity holds steady, avoiding the trap of fleeting fashions. A simple, well thought out design will likely be just as effective and beautiful years from now, while a design that leans on visual or interaction gimmicks can feel outdated much sooner.

Of course, simplicity is not the answer for everything. Sometimes, a touch of complexity can elevate a design, making it feel richer, more engaging, or even more delightful. But there is an art to adding depth without going overboard. Great designers find elegance in complexity by using it purposefully and sparingly, creating layers that enhance rather than distract.

The key is to balance complexity so it feels additive, not chaotic. Complex design should intrigue without being overwhelming, engaging users and inviting exploration. Adding complexity for the right reasons, a bit of depth here, a bit of flair there, can turn a functional experience into something that delights.

Knowing When to Embrace Complexity

Depth Over Density

Complexity works best when it creates depth, not density. Think about how layered navigation can help users explore more without feeling cluttered, or how a meaningful micro interaction can elevate the experience without demanding constant attention.

Purpose Driven Complexity

Adding complexity just to look fancy tends to backfire. But if a complex feature genuinely enhances the experience, like a customizable dashboard for a data heavy app, it is adding value that matters. The key is to question each addition: Does this truly enhance the user experience?

Creating Moments of Delight

Complexity has its place in those "Aha" moments where users discover a unique interaction or feature that delights them. When complexity creates joy, it has served its purpose beautifully. These moments do not overwhelm, they elevate.

Layered Navigation and Hidden Depths

Some of the best designs lead users down a layered path, revealing more options or features as they engage with the product. Complexity done right invites users to go further at their own pace, adding sophistication without overpowering simplicity. It is really about meeting users where they are.

Invisible Design: When the Best Design Is No Design at All

Not every design wants the spotlight, and not every element needs to show off. Some of the most powerful design choices are the ones users hardly notice. Invisible design works so seamlessly that it feels like second nature. The user does not even realize they are being guided. Invisible design is quiet competence. It makes a product feel obvious in the best way.

Invisible design means keeping things clear, accessible, and beautifully direct. It is about creating an experience where the user does not have to stop and think. They just move.

Guiding Without Forcing

Invisible design nudges users along the right path, providing just enough cues for them to intuitively know what to do. It is guidance without pressure.

Elegance in Utility

Invisible design can create a sense of calm and control for the user. A perfectly placed back button, a well labeled form field, a clear error state, these are small touches that make the experience feel effortless.

Letting the Content Shine

Sometimes a bold design can pull focus from what really matters, the content itself. Invisible design keeps the frame simple, letting the message be the hero. The experience becomes immersive rather than performative.

Seamless Interaction

Invisible design often hides the heavy lifting behind the scenes. Whether it is a smooth transition, a fast-loading experience, or an intuitive swipe, invisible design builds an experience that feels like magic because it simply works.

Navigating between simplicity and complexity is part intuition, part experience, and part restraint. Some designs call for subtlety, others for richness. Knowing when to flex each muscle is the mark of a mature designer.

The Editor's Eye: A Practical Cut List

If "editing" sounds subjective, that is because it can be. But great editorial judgment is not random taste. It is a repeatable way of choosing. The editor's eye is the ability to spot when the work is drifting away from its purpose, even if the drift looks beautiful.

Here is a practical cut list you can run on any design, feature, or flow. Use it like a checklist before you ship, and especially when your work feels bloated.

1. The Purpose Cut

Ask: If the user could only do one thing here, what should it be?

If the design does not make that obvious, simplify. Remove competing calls to action. Reduce choices. Create a clearer hierarchy. Editorial judgment starts with ruthless clarity about the purpose.

2. The Cognitive Load Cut

Ask: Where are we making the user hold too much in their head?

Look for long forms, multi-step decisions, unclear labels, hidden rules, and moments where the user must remember something from a prior screen. Then edit for relief. Break steps apart. Add progress cues. Make rules visible at the moment they matter.

3. The Trust Cut

Ask: What might make someone hesitate, worry, or abandon?

Trust is built in small details: error states, empty states, confirmations, reversibility, plain language. Sometimes the most important "feature" is not a feature at all. It is the feeling of safety. Editing for trust often means adding clarity while removing flourish.

4. The Friction Cut

Ask: What slows the user down without adding value?

Every extra tap, field, modal, and decision has a cost. If friction is intentional, it should protect the user from harm or confirm a meaningful choice. If it is accidental, cut it. Invisible design lives here.

5. The Maintainability Cut

Ask: Can the team support this six months from now?

Complexity is not just a user problem. It is a team problem. A design that requires constant special casing can drain engineering time, QA effort, and future iteration speed. Editorial judgment includes respect for what it takes to sustain the work.

6. The Narrative Cut

Ask: Does this tell one clear story, or three competing ones?

Design is storytelling through structure. When a screen tries to introduce a new value proposition, educate, upsell, and gather data, the story gets muddy. Editing is choosing the story that matters most for this moment.

If you want a single sentence to summarize editorial judgment, it is this: do not ask, "What can we add?" Ask, "What does this need to become obvious?" That question protects simplicity without worshiping it. It also allows complexity when it earns its place.

The Strength of Sacrifice

Sacrifice is difficult, but it often leads to sharper, more effective design. When you refine down to the essentials, each remaining element has room to shine, without distraction or dilution. Trimming the design to focus on the must haves ensures the product has clarity and purpose.

This restraint is not about leaving things out for the sake of simplicity. It is about making sure that everything included has an intentional role to play. Sacrifice becomes a tool for honing the design, keeping it aligned with the product's mission and amplifying its impact.

As a designer, you are naturally inclined toward creativity and innovation, and the ideas that excite you most are often the boldest, most unconventional ones. Sometimes the wildest ideas do not fit within the current product's goals or constraints. Knowing when to set aside those ideas, without discarding them forever, is an under discussed skill in product design, and it is one that separates a good designer from a great one.

Some ideas are not right for the immediate project, but they are not bad ideas. They might be exactly what is needed later, in a future version of the product or a different context altogether. Recognizing when to defer these ideas is key.

A Simple Framework for Deciding What Stays

Assess Immediate Relevance

Does the idea address a current user pain point, or is it more aspirational? Sometimes a feature is wonderful, but better suited for later, when the core product is more established and there is room to experiment.

Consider Feasibility and Cost

An idea you love might be brilliant but technically complex or time intensive. If resources are limited, let it go for now and focus on what can be implemented smoothly and reliably, without compromising quality.

Evaluate Against Business Priorities

Product design serves the user and the business. Some beloved features might add beauty or whimsy but do not deliver on business outcomes. Understanding how your team defines value helps guide what you choose to keep or defer.

One way to soften the difficulty of cutting an idea is to adopt a "not now, maybe later" mindset. Capture these ideas in a backlog or design library, a creative vault. This does not mean endlessly accumulating ideas. It means giving them a place to breathe. Often, something you shelved finds new relevance later as the product, market, or technology evolves.

MVP as an Editorial Practice

In today's tech landscape, the concept of an MVP, Minimum Viable Product, is fundamental. The MVP distills a product down to its essential features, enough to satisfy early adopters and provide feedback for future development. But MVP does not mean ugly or uninspired. It means focused. Within an MVP, every element has a purpose, and nothing is superfluous.

For designers, working within the constraints of an MVP can be liberating. It gives clear boundaries, freeing you from the temptation to over design. An MVP's value lies in its clarity, in its ability to communicate what the product is at its essence.

The MVP also builds discipline, teaching you to spot must-haves versus nice to haves quickly and decisively. The MVP is not just a minimal version of your product. It can be the most refined version, allowing every element that makes the cut to be meaningful.

Sometimes the business side will prioritize functionality over a "wow" factor, opting for basic over bold. This can be a struggle. It is also part of the craft, aligning with business needs while still finding opportunities for care, delight, and craft. If you feel creatively constrained, do not be afraid to let that energy flow into passion projects outside of work. Having outlets for uninhibited expression can make you a stronger designer within a structured environment. If you feel tension here, be honest with yourself about what you need and how you recharge.

Design is not about adding everything you can. It is about knowing when to stop. Sometimes the greatest impact lies in what you are willing to let go. The journey from idea to reality is one of constant refinement, where everything you create does not always make it to the final product.

Carry This Forward

Editing is a form of leadership. It is how you protect the user from clutter, the team from noise, and the product from becoming everything at once. Carry forward the willingness to choose, to sacrifice the clever extra in service of the essential thing that truly helps someone. Let your restraint be generous. Let your decisions make space for clarity. When you practice the art of the edit, you do not just ship more focused work. You become a more intentional designer.

Chapter Reflection

Reflection Prompts

- Where in your current work are you adding more than is truly needed? What would happen if you removed one thing?
- What is the hardest part of editing for you; letting go of your own ideas, navigating constraints, or aligning with others?
- What is one idea, feature, or flourish you love that belongs in a "not now" vault instead of this release?

- Where could invisible design serve your users better by reducing friction, attention, or effort?

Artistic Expression

Create a two-panel piece called "Everything" and "What Remains." On one side, sketch or collage a crowded screen, flow, or concept filled with every idea you could include. On the other side, redraw it with only what truly needs to be there. Add small symbols next to what you cut and what you kept. Notice which version feels more honest.

Soulprint Words

Choose 3 to 5 words that capture what strengthened in you around editorial agency.

Examples: Focused, Discerning, Editor, Intentional, Grounded, Precise

Soulprint Thread

Your edge becomes visible here. Carry forward a mark or phrase that reminds you, "I know when to stop so what matters can shine." In your Soulprint, this is your contour line, the boundary that clarifies your work and protects its purpose.

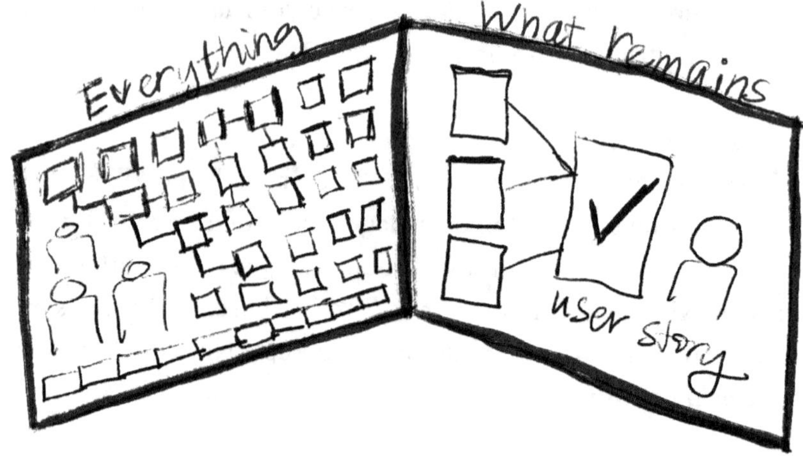

Everything
What remains
user story

| 8 |

Metabolizing Tension

You have explored how restraint, simplicity, and sacrifice shape what actually ships, and why focus is one of your most powerful tools. Now comes the moment where your work leaves your hands and enters a room. This is the shared space of design, where collaboration introduces tension, feedback, competing priorities, and real constraints. What began as something personal becomes something collective.

Collaboration is not just about working together. It is about learning how to hold your work steady while other people shape it. Priorities collide. Opinions multiply. Constraints tighten. Feedback lands imperfectly. Learn to work inside that tension, not eliminate it. Tension is not a sign that something has gone wrong; it is a signal that something matters.

Metabolizing Feedback: Turning Noise into Nutrients

Feedback is not just information. It is energy. It arrives with tone, timing, context, and sometimes an entire backstory you are not aware of. Some feedback nourishes the work immediately. Some feedback is hard to digest. Some feedback is not even about the work, it is about fear, identity, authority, or politics.

Metabolizing feedback means you do not swallow everything whole, and you do not spit everything out. You process it. You slow down long enough to extract what is useful, release what is not, and transform the rest into something that strengthens the work.

This is a design skill and a leadership skill. It is what allows you to stay open without becoming porous, to stay confident without becoming defensive, and to keep moving without losing yourself.

The Feedback Metabolism Loop

When feedback hits, try moving through these steps. Not perfectly. Just intentionally.

1. Receive: Let it land without reacting

Your nervous system will want to do something fast. Defend. Explain. Fix. Freeze. Please.

Instead, practice receiving first. Breathe. Take notes. Ask yourself: What did they actually say? What are they feeling? What are they protecting?

Receiving is not agreeing. It is making space.

2. Decode: Find the need under the note

Most feedback is a clue, not a command. "I do not like it" might mean "I do not understand it." "This feels risky" might mean "I cannot predict the outcome." "Can we make it simpler" might mean "I am afraid users will fail."

Decode by asking:

- What problem are you worried about?
- What user are you imagining?
- What would success look like to you?
- Can you show me where this breaks down?

You are not arguing. You are translating.

3. Sort: Separate signal from style

Not all feedback is equal, and not all feedback should steer the work.

Listen for three types:

- **Direction feedback:** Are we solving the right problem?
- **Execution feedback:** Is it clear, usable, accessible, technically feasible?
- **Taste feedback:** Do they simply prefer a different visual or pattern?

All three can be useful, but only the first two should drive strategy. Taste can inform craft, but it should not replace purpose.

4. Choose: Decide what you will do with it

Metabolizing feedback requires judgment. The goal is not to incorporate everything. The goal is to incorporate what improves the work.

When you choose, be explicit:

- **Adopt:** This improves the design and aligns with goals.

- **Adapt:** The underlying concern is valid, but we will solve it differently.
- **Defer:** Good insight, wrong moment. Capture it.
- **Decline:** This pulls us away from user needs or intent.

The most mature response to feedback is not always change. Sometimes it is clarity.

5. Integrate: Make the work stronger, not noisier

When you integrate feedback, do it with intention. Do not patchwork the design until it becomes a Frankenstein of approvals.

A helpful check:

- Does this change improve comprehension?
- Does it reduce risk or failure for users?
- Does it align with the decision we are making right now?

Integration is where collaboration becomes craft.

6. Close the loop: Reflect it back to the room

People trust the process when they see their input was heard, even if it was not followed exactly.

Close the loop with language like:

- "I heard your concern about X. We addressed it by doing Y."
- "We tried the suggestion and it introduced Z tradeoff, so we adapted it."
- "We are deferring this idea until after we validate the core flow."

Closing the loop turns feedback into partnership instead of a drive-by critique. Before we talk about patterns and phrases, we need to talk about what feedback actually feels like inside the body of a designer. Because collaboration doesn't just test the work; it tests us.

The Quiet Weight of Collaboration

Collaboration is where design becomes real. It is also where it becomes vulnerable.

We often talk about feedback as if it is neutral information, something to be received, processed, and applied. But feedback never arrives empty handed. It comes wrapped in tone, timing, power, history, and expectation. It lands in a body that has already invested time, care, and belief into the work.

That matters.

For designers especially, feedback is not just about the artifact. It can feel like a referendum on judgment, taste, intelligence, or worth. Even seasoned designers who know better still feel it in their chest when a room goes quiet, when a leader frowns, when someone says, "I'm not sure about this" without saying why.

We do not experience feedback only intellectually. We experience it physically.

A tightening. A rush of heat. A sudden urge to explain, justify, or retreat.

That reaction is human. It does not mean you are fragile. It means you care.

The goal is not to stop feeling that reaction. The goal is to learn how to stay present inside it.

Because collaboration is not just a design skill. It is an emotional one.

When Feedback Isn't Really About the Work

One of the hardest lessons I learned as a design leader was this: not all feedback is actually about design.

Sometimes feedback is about fear. Fear of change. Fear of risk. Fear of being accountable for a decision that might not work.

Sometimes it is about control. A need to feel relevant, powerful, or included.

Sometimes it is about identity. A stakeholder protecting what they built. A leader protecting their vision. A team protecting their way of working.

And sometimes, feedback is simply about timing. The room is tired. The pressure is high. The stakes feel personal.

If you treat all feedback as a literal instruction, you will exhaust yourself and dilute your work. If you treat all feedback as an attack, you will harden and disengage.

The work is in discernment.

You are not here to absorb everything. You are here to understand what is underneath it.

This is where designers often struggle. Not because they lack skill, but because they have not been taught how to separate signal from emotional noise without shutting down.

Power Changes the Shape of Feedback

Feedback does not exist on a flat plane. It is shaped by power.

A suggestion from a peer lands differently than a comment from a senior leader. Feedback in a public forum feels different than feedback in a one on one. Silence from someone with authority can feel louder than critique from someone without it.

Designers notice these dynamics even when no one names them.

When someone with power gives vague feedback, it can freeze a team. People try to read between the lines. They redesign for approval instead of clarity. The work becomes cautious.

When feedback feels performative or political, designers can start to second guess their instincts. Over time, this erodes confidence, not because the designer lacks talent, but because the environment rewards safety over truth.

It is important to say this plainly: navigating power is part of the job. That does not mean accepting poor behavior. It means understanding the system you are operating in so you can choose how to respond without losing yourself.

You are allowed to ask for clarity.

You are allowed to slow the conversation down.

You are allowed to name tradeoffs.

Doing so is not being difficult. It is being responsible.

The Cost of Shrinking

Many designers respond to tension by shrinking.

They simplify ideas too early. They remove the bold part. They preemptively compromise to avoid friction. They translate their thinking into whatever language feels safest for the room.

At first, this can feel like collaboration. Over time, it becomes self-erasure.

I have seen brilliant designers slowly go quiet. Not because they ran out of ideas, but because they stopped believing those ideas would be held with care.

This is one of the most dangerous outcomes of poorly metabolized collaboration. Not that the work gets worse, but that the designer disconnects from their own conviction.

The irony is that teams need designers most when things feel uncertain. When problems are ambiguous. When the room is stuck.

That is when your perspective matters most.

Staying whole in collaboration does not mean being inflexible. It means staying anchored to intent even as form evolves.

How to Stay Present Without Absorbing Everything

There is a difference between openness and permeability.

Openness means you are willing to listen, to be influenced, to change your mind when the work improves.

Permeability means everything passes through you without filter, including anxiety, projection, and fear that are not yours to carry.

Metabolizing tension requires boundaries.

Here are a few practices that help designers stay present without absorbing everything in the room.

Name What You're Solving For

Before feedback begins, ground the room.

Say what this work is doing right now. Is it exploring? Is it deciding? Is it validating? Is it reducing risk?

When people know the purpose, their feedback becomes more useful. When they do not, they often project their own expectations onto the work.

This small act protects both the design and your nervous system.

Slow Down the Reaction

You do not need to respond to feedback immediately.

It is okay to say, "Let me sit with that."

It is okay to say, "I want to make sure I understand what you're concerned about."

It is okay to take notes and come back later.

Speed is often confused with competence. In reality, thoughtful pacing builds trust.

Ask for the Why, Not the Fix

When feedback comes in the form of a solution, gently pull it back to the underlying concern.

- "What problem are you worried this creates?"
- "What would success look like here?"
- "Where do you see this breaking for users?"

This keeps you in the role of designer, not order taker.

Decide What Belongs to You

Not all discomfort is yours to process.

Some tension belongs to the system. Some belongs to leadership misalignment. Some belongs to a timeline that is too aggressive.

Your job is not to internalize everything. Your job is to design within reality without becoming consumed by it.

Collaboration Without Consensus

One of the most damaging myths in product teams is that alignment means agreement.

It does not.

Alignment means shared understanding of the decision being made, the tradeoffs involved, and the direction forward, even if not everyone prefers it.

Designers often feel pressure to make everyone happy. This leads to watered down solutions that satisfy no one and serve users poorly.

You are allowed to move forward with clarity even when consensus is imperfect.

In fact, that is often your responsibility.

When you can articulate why a decision was made, what was considered, and what will be revisited later, you create psychological safety without stalling progress.

Clarity is a form of care.

The Designer's Inner Line

Every designer has an internal line. A point beyond which compromise becomes misalignment.

Learning where that line is takes time. Crossing it once or twice usually teaches you quickly.

That line is not about ego. It is about integrity.

It might sound like:

- "This no longer serves the user."
- "This erases the core insight."
- "This turns the experience into something I cannot stand behind."

Honoring that line does not mean refusing collaboration. It means naming when the work is drifting away from its purpose.

You will not always be able to hold that line. Power dynamics are real. Constraints are real. Sometimes the decision will move forward without you.

But knowing where the line is changes how you show up. You stop confusing compliance with professionalism. You start making intentional choices about when to push, when to adapt, and when to let go.

Leaving the Room with Yourself Intact

At the end of a hard meeting, ask yourself:

- Did I listen?
- Did I advocate?
- Did I stay connected to my intent?

If the answer is yes, even if the outcome was not what you hoped for, you did your job.

Collaboration is not about winning; it is about participating with integrity.

Some of the most meaningful design moments do not result in shipped features. They result in reframed conversations. Slower decisions. A team that sees a problem differently than they did before.

That is impact too.

You will not always get credit for it. But you will feel it.

And over time, others will too.

Because designers who can stay open without dissolving, who can metabolize tension without becoming rigid or porous, become anchors in complex systems.

They become trusted.

They become steady.

They become leaders.

Feedback Patterns: What People Say vs What They Mean

Here is the tricky part about feedback in a room: people rarely give you clean feedback. They give you their reaction. They give you their anxiety. They give you the version that is safest for them to say out loud. Your job is to metabolize what they offer and uncover what the work actually needs.

Below are common phrases you will hear, and what they often mean underneath. This is not to make you cynical. It is to make you precise.

"I don't like it."

Often means: I do not understand it, I do not trust it, or it does not match what I expected.

Response: "What part feels unclear or risky? What did you expect to happen instead?"

"Can we make it simpler?"

Often means: I am worried users will fail, or I am worried support costs will rise.

Response: "What would a user struggle with? What failure are you most concerned about?"

"This feels off brand."

Often means: This threatens consistency, trust, or identity.

Response: "What specific brand principle is being broken? Is it voice, visual system, interaction tone, or something else?"

"Do we really need this?"

Often means: I am protecting timeline, scope, or risk.

Response: "If we cut it, what is the user impact? If we keep it, what cost are we accepting?"

"I'm not sure users want this."

Often means: I do not want to own the decision, or I am uneasy without data.

Response: "What signal would help you feel confident? What can we test quickly?"

"We have to ship next week."

Often means: We need a clear tradeoff, not a perfect solution.

Response: "What is the smallest version that preserves user value and reduces risk?"

When you respond this way, you are not being difficult. You are elevating the conversation. You are helping the room shift from preference to purpose. This is the difference between collaboration that drains you and collaboration that sharpens you.

Also, a quiet truth: metabolizing feedback gets easier when you stop treating every comment as a referendum on your ability. Some feedback is about the work. Some is about the context. Some is about the person delivering it. Your steadiness comes from remembering that your job is not to be liked, it is to be useful.

Where Feedback Gets Hard to Digest

Metabolizing feedback is easy when it is thoughtful and specific. It is harder when the room is complicated.

1. When Stakeholders Disagree

Sometimes feedback conflicts because priorities conflict. One leader wants growth. Another wants trust. Someone is worried about risk. Someone is protecting their team. Someone is protecting their reputation.

When feedback conflicts, do not try to satisfy everyone. Instead, name the tradeoff:

If we optimize for speed, we may lose clarity. If we optimize for clarity, we may slow onboarding. Which cost can we afford right now?

Alignment is not a feeling. It is a decision.

2. When Engineering Constraints Collide with User Needs

Constraints can feel like a wall. But constraints often hold information you do not have. Performance, maintainability, architecture, reliability. These are user needs too, just delayed.

Invite engineering into the conversation earlier. Ask for options. Create a good, better, best set of paths. Design is not just the ideal. It is the best possible within reality.

3. When Feedback Is Vague or Performative

Not all feedback is designed to help. Sometimes it is theater. Sometimes it is someone trying to control the outcome without owning the decision.

When feedback is vague, make it concrete:

- "What are you worried will happen?"
- "Where specifically does this break?"
- "What would you want to measure?"

Clarity is your best tool when feedback is foggy.

4. When Brand and Product Priorities Conflict

Brand wants consistency, craft, and distinctiveness. Product wants speed, conversion, and clear behavior. Both are right.

Translate brand into user value. Identify what must be consistent and where there is flexibility. The best products feel usable and unmistakably themselves.

5. When Timelines Force Tradeoffs

Deadlines compress everything. Teams cut corners. They skip research. They ship a compromise and tell themselves they will return later.

Sometimes you will return later. Sometimes you will not.

Make tradeoffs explicit. Protect the highest impact moments. Keep a visible backlog of intentional debt. A rushed decision is still a decision. Treat it with care.

The North Star vs. What Ships Tomorrow

There is the dream design and the version that can realistically ship soon. Strong designers hold both. The trap is thinking you must choose one or the other. You do not.

The North Star is not a fantasy. It is a direction. Shipping is not selling out. It is delivering value now.

The skill is sequencing. Ask: What piece of the vision can we introduce now that moves us in the right direction? That might mean shipping the simplest version of a new pattern, then layering sophistication over time. It might mean building the foundation first, then adding delight later. It might mean shipping a safe baseline so you can earn the trust to try something bolder next.

This is not compromise as defeat. It is compromise as strategy.

Do Not Forget the Joy

There is a particular joy in building with others, in watching ideas change shape through conversation and shared effort. That joy sustains designers through complexity and fatigue.

A team affectionately called our ideation sessions "Heideation." It reflects a shared trust that the process will be messy, exploratory, and worth it. On the other side, we usually find clarity and confidence we could not have reached alone.

I have learned that I need solo time before collaboration. Without it, my thinking can become overly influenced by the room. Knowing when to step back and when to step in took years to understand. Every designer must find their own rhythm between solitude and collaboration.

Great collaboration does not erase individuality. It depends on it.

Carry This Forward

Metabolizing feedback is how you stay open without losing your center. Carry forward the ability to receive input without absorbing it whole, to translate reactions into needs, and to make decisions that protect the work instead of patching it. Tension is not something to fear. It is proof the work matters to more than one person. Your job is to process it with integrity, turn it into clarity, and keep building.

Chapter Reflection

Reflection Prompts

- What is your default reaction to feedback: defend, explain, fix, freeze, or please? What would it look like to receive first?
- Think of the last piece of feedback that stung. What was the need underneath it?
- Where are you currently absorbing too much input, and where are you dismissing too quickly?
- What is one tradeoff you are avoiding naming out loud? What changes if you name it clearly?

Artistic Expression

Create a three-stage visual called **"Raw, Processed, Integrated."**

Stage 1: draw feedback as it first arrives, chaotic, sharp, noisy.

Stage 2: draw what happens inside you when you metabolize it, filtering, sorting, extracting.

Stage 3: draw the final form, where the work is stronger and clearer.

Add one small symbol that stays consistent across all three stages. That symbol is your center.

Soulprint Words

Choose 3 to 5 words that capture what this chapter strengthened in you.

Examples: *Steady, Translator, Grounded, Resilient, Clear, Diplomatic*

Soulprint Thread

Your internal compass becomes visible here. Carry forward a symbol or phrase that reminds you, "I can metabolize feedback without losing myself." In your Soulprint, this becomes the thread that keeps you centered while you build with others.

| 9 |

Data With Soul

Conviction grows stronger when it's informed. Let's dig into how data deepens design; how numbers and intuition can work together to create meaning. Think of data as your ally, something to embrace, understand, and use as a guide. Data gives us clarity. It reveals patterns, highlights gaps, and uncovers opportunities we might otherwise miss. It's about more than just crunching numbers; it's about understanding the stories behind the data points and how they shape user experiences.

Metrics are like our compass in design, they show us where we are, where we're headed, and how to navigate to the best possible user experience. They're not just numbers; they're insights into what users love, what frustrates them, and where we can make meaningful changes. By embracing metrics, we're not only improving our designs but also our understanding of how people interact with what we create.

Embracing Data in Design

First off, get to know how your product is actually used by customers. This means observing users, conducting qualitative studies, and diving deep into stats about your product's performance after it hits the market. The most important tools for you to have access to is behavioral data and access to people who use your product. If you don't have access, ask your leadership for what you need to be successful. The best way to go about this is outlining questions you have about how customers interact with the product. A helpful way to start is to turn your curiosity into a repeatable set of questions, then map each question to a specific metric, event, or dataset you can request. You may need to engage in new partners to provide this data, but usually it takes partners in customer service and data to provide the data you need. Highlight the iterative process of design and measurement.

Practical Sampling of Designer-Friendly Questions + the Data to Ask for

"Are people succeeding or getting stuck?"
Ask:

- Where do users drop off in the critical flow?
- What step has the biggest time sink?
- Where are users looping or retrying?

Data to request:

- Funnel conversion by step (view → start → complete)
- Step-level drop-off rates
- Time-to-complete by step (median + p90)
- Error rates by step (validation errors, API errors)

- Rage clicks / repeated taps (if your tooling supports it)
- Retry counts (submit attempts, resends)

"What are people doing right before they bounce or abandon?"

Ask:

- What's the last action before exit?
- How often are users hitting back?
- Are they bouncing because they're confused or because they're done?

Data to request:

- Back button events (hardware back + in-app back)
- Exit pages / last screen before session ends
- Bounce rate (for web) and short sessions (for app)
- Session duration distribution (median + p90)
- Scroll depth and "did they reach the CTA"
- Time on page/screen before exit

"Where do users come from, and does that change what they need?"

Ask:

- Are marketing users behaving differently than organic users?
- Do users from email campaigns convert worse because the landing context is wrong?
- Which entry points produce the healthiest retention?

Data to request:

- Acquisition source / channel (UTM, referrer, campaign)
- Landing page or entry screen
- New vs returning user behavior by channel
- Conversion and retention by source
- Geo + language breakdown (if relevant)

"Is the experience consistent across devices and contexts?"
Ask:

- Is mobile conversion lower than desktop because the flow is harder?
- Are certain browsers or OS versions breaking the experience?
- Are load times killing completion?

Data to request:

- Device type, OS version, browser version
- Screen size / viewport
- Performance metrics: page load, time to interactive, API latency
- Crash rate and crash-free sessions (mobile)
- Network type (WIFI vs cellular), if available

"What do users search for, and are we helping them find it?"
Ask:

- What are the most common searches, and do they lead to success?

- Are users searching for things we don't have or don't label well?
- Are filters being used or ignored?

Data to request:

- Search queries (top terms + zero-result searches)
- Search-to-click rate
- Search-to-conversion rate
- Filter usage and most-used facets
- "No results" events and follow-up behavior

"Which parts of the UI are actually used?"
Ask:

- Are users using the feature we built, or skipping it?
- Which actions are used repeatedly (habit-forming)?
- What's getting ignored?

Data to request:

- Feature adoption (users who use it / eligible users)
- Frequency of use per user (weekly/monthly)
- Click/tap heatmaps (if you have them)
- Interaction events for key UI components
- Path analysis (common sequences)

"Are we building trust and reducing support burden?"
Ask:

- What do customers complain about most?
- Which UX issues drive tickets, refunds, or escalations?

- Where do people get anxious or confused?

Data to request:

- Support ticket tags and top themes
- Contact rate per active user
- Refund/cancel reasons (if applicable)
- Chat transcript themes or macro usage
- NPS/CSAT verbatims mapped to product area (when possible)

"Are we improving over time?"
Ask:

- Did the redesign actually move the needle?
- Did we improve one metric but harm another?
- Who benefited and who didn't?

Data to request:

- Pre/post metrics for the same flow
- Cohort retention (D1, D7, D30) by segment
- A/B test results, experiment readouts
- Guardrail metrics (errors, support tickets, churn)
- Segmented results (new users vs power users)

Getting Started with Data

Metrics give you hard data to base your design decisions on, instead of just going with your gut. With a clear benchmark, you can track how your design is doing over time and see if you're making progress. By linking design metrics to business goals (like how many users stick around or make a purchase), you can show the real impact of good design. Knowing what works and what doesn't can spark new ideas for you and lead to innovative solutions.

When starting a project, begin by pinpointing a key success metric that you aim to improve or create. This metric becomes your guiding objective: a specific target to enhance. Establish baseline metrics for your product and consistently monitor them. Alongside product success metrics, designers should establish baseline behavioral metrics that reveal how people actually move through the experience; where they pause, hesitate, backtrack, or succeed. These metrics complement what product managers track, but they serve a different purpose: understanding behavior, not just outcomes. Collaborate closely with product managers to define these metrics. After each release, assess whether there has been improvement. It also helps in prioritizing what aspects require attention and refinement. I've observed instances where meticulously designed elements received minimal user engagement. While this may signal a need for further exploration, it underscores the importance of understanding user priorities and preferences. Numbers don't lie!

Types of Metrics and Tools

There are many types of metrics in product design, broadly falling into quantitative and qualitative categories. Both matter, and neither is sufficient on its own. Strong design decisions come from understanding *what* users do and *why* they do it.

Quantitative metrics tell you what is happening at scale. Qualitative insights help you understand what it feels like and why it's happening. Designers need access to both to design responsibly and effectively.

Just as importantly, designers should build close relationships with teams who sit closest to customer pain. Customer service, support, and success teams carry enormous insight into where users struggle, often long before those issues show up clearly in dashboards. Good design can materially reduce support volume, repeat contacts, and customer frustration, which is why understanding these signals is not just helpful, it is strategic.

Below is a structured way to think about metrics, tools, and studies across the design lifecycle.

Quantitative Metrics (What's Happening)

These metrics are behavioral, measurable, and scalable. They help you see patterns across large groups of users.

Common quantitative metrics

- Conversion rates (sign-up, purchase, activation)
- Task success and completion rates
- Drop-off and abandonment points
- Time on page or time in flow
- Back-button usage or rage clicks
- Feature adoption and usage frequency
- Retention, repeat visits, and churn
- Load times and performance indicators
- Error rates and failed actions
- Funnel progression and falloff

Why this matters for designers

Quantitative data shows where friction exists, where attention drops, and where behavior diverges from intent. It helps you prioritize which problems deserve deeper exploration, not guess based on opinion or volume alone.

Behavioral Metrics (How People Actually Move)

Behavioral metrics sit between qualitative and quantitative. They show real user behavior in context, often revealing issues users cannot articulate.

Examples

- Heatmaps (clicks, scroll depth, attention)
- Session recordings
- Form interaction patterns
- Hesitation, looping, or repeated actions
- Navigation paths and unexpected detours

Why this matters for designers

Behavioral data tells the story behind the numbers. You can see confusion, friction, and intent unfold moment by moment. It is especially powerful for validating usability assumptions and spotting breakdowns that analytics alone cannot explain.

Qualitative Metrics (Why It's Happening)

Qualitative data captures human experience. It explains motivation, emotion, and perception.

Examples

- User interviews
- Usability testing
- Open-ended survey responses
- Diary studies
- Customer feedback and complaints
- Support tickets and call transcripts

Why this matters for designers

Qualitative insights reveal emotional friction, unmet needs, and mental models. This is where you understand *why* users behave the way they do and how design choices shape trust, confidence, and satisfaction.

Emotional Design and Brand Signals

Design choices influence how users feel, not just how they function.

What to look for

- Emotional reactions in interviews
- Language users use to describe the product
- Trust, delight, frustration, or hesitation
- Perceived clarity or confidence

This work is strongest when product and brand teams collaborate. Shared understanding of brand essence, tone, and emotional goals helps create experiences that feel coherent and intentional, deepening customer connection over time.

Usability, Accessibility, and Support Signals

Usability and accessibility are foundational, not optional.

Metrics and inputs

- Task success rates
- Accessibility audits and compliance checks
- Error frequency

- Support volume related to specific flows
- Repeat contact rates for the same issue

Customer service and support teams are especially important partners here. They see patterns of confusion and frustration early, and good design can dramatically reduce their workload while improving the customer experience.

Performance and Experimentation Metrics

Design does not live apart from system performance.
Examples

- Page load time
- App responsiveness
- Error states and failure rates
- A/B testing results
- Multivariate testing outcomes

These metrics help ensure that a beautifully designed experience actually works reliably in the real world.

Pre-Launch vs. Post-Launch Studies

Design research should change as the product moves through its lifecycle.
Before launch

- Generative research
- User interviews
- Concept testing

- Usability testing
- Accessibility reviews
- Prototype validation

After launch

- Behavioral analytics
- Heatmaps and session recordings
- A/B testing
- Customer feedback analysis
- Support ticket review
- Long-term engagement tracking

Before launch, you are reducing risk. After launch, you are learning from reality.

Long-Term Engagement and Value

Some of the most important signals emerge over time.
Examples

- Repeat usage
- Feature adoption
- Habit formation
- Referrals and advocacy
- Retention across cohorts

These metrics help you understand whether your design delivers sustained value, not just short-term success.

Bringing It Together

No single metric tells the full story. Design maturity comes from connecting behavior, emotion, and outcomes.

When designers combine quantitative data, behavioral insight, qualitative feedback, and frontline customer perspectives, they move from designing screens to shaping systems that genuinely serve people.

Collaborating with Data Experts

Throughout my career, I've partnered closely with data experts in my organization for a couple of reasons. Firstly, these partners have a unique way of thinking about numbers and patterns that can offer fresh perspectives. Secondly, I trust their depth of analysis more than my own dashboards or interpretations. So, designers, consider data managers as your best allies. Your job is to ask good questions and work with them to gather the data that will help you build the best product possible.

I would also put key research folks in this camp of data experts; it's just a different type of data. There's a common misconception that integrating qualitative data into your design practice requires a dedicated research team. While having a research team is great, it shouldn't replace these hands-on methods. In my opinion, designers develop a stronger sense of empathy and care more about their users when they engage in this research and data analysis themselves. So, don't shy away from it — dive straight in!

It's easy to misinterpret data or use it in ways that lead to poor decisions, this is why data partners are important to have as a starred or favorite contact.

Balancing Data in Design

I like to think about knowing your users alongside knowing the numbers. Metrics tell you what is happening, but they rarely explain why. Being data-informed means using data as an input, not a verdict. It means letting numbers sharpen your questions, validate patterns, and surface risk, while still relying on experience, intuition, and context to make sense of what those numbers actually mean.

A data-informed designer looks beyond dashboards. They pair analytics with customer feedback, usability insights, support conversations, and industry context before deciding what action to take. I've seen teams notice a behavior shift in the data, then validate it through customer surveys, interviews, or usability studies before making a change. That balance is where strong decisions live.

A strictly data-driven approach, on the other hand, treats metrics as absolute truth. When a team optimizes a single number — click-through rate, conversion, time on page — without considering qualitative feedback, long-term impact, or strategic intent, they risk making decisions that are technically optimized but experientially wrong. That is not objectivity. It is tunnel vision.

Being data-informed is about responsibility. It is about respecting what data can reveal without outsourcing judgment to it. The goal is not to remove human insight from decision-making, but to strengthen it.

I'll add that empirical certainty is a bit of a myth. We crave clear answers and definitive proof, but in reality, everything is more nuanced. Data, like design, always leaves room for interpretation. It's like trying to capture the ocean in a bucket — you may gather a sample, but it will never be the whole story. Embracing that uncertainty keeps us curious, open, and willing to learn. Design plays a critical role here by turning numbers into narratives and helping teams understand not just what changed, but why it matters.

Design systems are complex, with many interdependent parts, which makes it difficult to isolate exactly what caused a shift. Gathering and analyzing data takes time and can be expensive, and not every question is worth answering exhaustively. Do not let the pursuit of perfect data prevent you from learning. Follow your curiosity. As you deepen your understanding of users, better questions will naturally emerge.

Use both quantitative and qualitative insights to show stakeholders the value of your design work. When you maintain this balance, your designs are not only effective, but meaningful. Data does not replace empathy. It reinforces it. And when integrated thoughtfully, it helps you design with greater clarity, confidence, and care.

Best Practices for Using Metrics

Once you've set your success and behavioral metric targets, the next step is regularly checking in on your data. It's like reading the story of how people are using your product; what's working well, where they might be getting stuck, and where there's room to make things even better.

Based on what you learn from the data, it's all about iterating your designs. This isn't just a one-time thing; it's an ongoing process of tweaking and refining based on what the metrics tell you. Maybe it's adjusting the layout to make it more intuitive, testing out different features to see what users respond to best, or fine-tuning the wording to make it more engaging or understandable.

I encourage you to share your findings with key partners. Whether it's your team, managers, or stakeholders, explaining how your design decisions connect with the metrics helps everyone understand the impact of what you're doing. It's about painting a clear picture of how design isn't just about making things look good — it's about creating experiences that truly resonate with users and drive positive outcomes.

In essence, metrics are your guideposts in the design journey, helping you navigate toward a product that not only meets but exceeds user expectations.

Accessibility: The Metric That Refuses to Behave

Some of the most important outcomes in design are the hardest to measure. Accessibility is one of the most meaningful.

Unlike page load time or conversion rate, accessibility does not resolve neatly into a single number. It is situational, contextual, and deeply human. What works for one person may not work for another. What works for the same person one day may fail them the next, depending on fatigue, environment, assistive technology, or cognitive load. Accessibility lives at the intersection of bodies, tools, systems, and moments. That complexity makes it uncomfortable for teams that want certainty.

I have watched accessibility become a debate framed around ROI, timelines, and edge cases. I have seen teams ask for proof before they offer consideration. And I have seen designers struggle to advocate for users whose experiences do not show up clearly in dashboards or weekly reports.

This is where judgment matters.

Accessibility exposes the limits of being purely data driven. You will not always be able to justify it with a clean metric. You will not always be able to A/B test dignity, independence, or inclusion. That does not make accessibility optional; it makes it a design responsibility.

When teams invest in accessibility thoughtfully, the impact is real, even if it arrives unevenly. Users encounter fewer barriers, and products feel calmer and more predictable. Systems become more resilient. Often, accessibility improvements quietly improve usability for everyone, clearer hierarchy, better focus states, more intentional flows. The benefits compound over time, not always in the same sprint they are introduced.

Accessibility also changes teams.

Designing with accessibility in mind forces clarity. It exposes assumptions. It slows teams down in productive ways. It pushes designers to think beyond the "default" user and ask who the system quietly excludes. That mindset carries into other decisions, language choices, error handling, onboarding, trust.

It is worth naming something else plainly. Accessibility is not a widget.

Adding paid overlays or one-click accessibility tools is often a signal that a team is optimizing for optics rather than experience. Many people who rely on assistive technology already have system-level tools configured to meet their needs. Forcing them to adapt again, site by site, product by product, creates friction, not inclusion. Watching real users navigate these layers makes this painfully obvious.

Accessible design is not about quick fixes. It is about respecting how people actually move through the world.

You may not always be able to prove the value of accessibility in the way leadership asks for. You may still need to advocate without perfect data. That is part of the work. This is where design becomes more than optimization. It becomes stewardship.

Accessibility reminds us that not everything meaningful can be measured cleanly, and not everything measurable is meaningful. Choosing to design for inclusion is not a detour from good business. It is a long-term investment in trust, usability, and humanity.

Carry This Forward

When data meets empathy, insight becomes wisdom. Carry forward your ability to see metrics not as measures of output, but as reflections of impact, the story behind the numbers.

Chapter Reflection

Reflection Prompts

- How do you currently use data to shape your design decisions?
- When has data clarified your path — and when has it clouded your creative instinct?
- What kind of data feels alive to you? What kind feels disconnected from your practice?
- How might you turn data into a dialogue, not just a report?

Artistic Expression

Create a visual language for data that feels soulful. Try drawing data not as charts, but as feelings, moments, or stories. What would a "pulse of trust" look like? A "metric of joy"? Play with metaphor: vines instead of bar graphs, weather instead of KPIs.

Soulprint Words

Choose 3–5 words that capture what the data topics has unlocked in you.

Examples: *Interpreter, Meaning-maker, Balanced, Insightful, Grounded*

Soulprint Thread

Data understanding reveals your ability to *blend rigor with resonance.*

Carry forward a symbol or phrase that reminds you:
"Data is not the answer; it's an input to help us ask better questions."

Culture is the Operating System

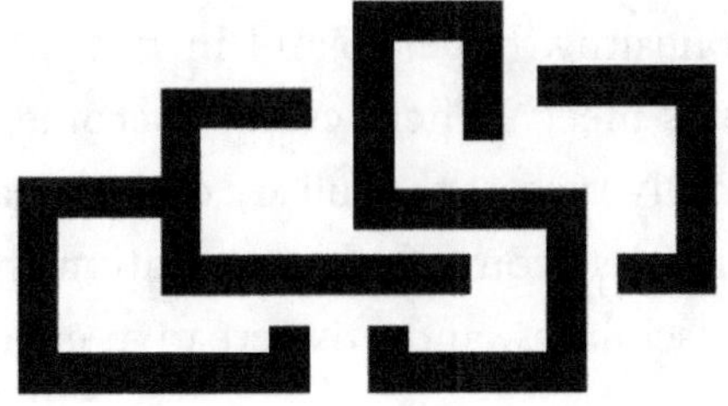

How Care, Clarity, and Rhythm Scale Design

Design teams don't run on tools alone. They run on care, clarity, and rhythm. Data can tell you what is happening, where work slows down, where users struggle, and where teams are overloaded. Culture determines how that data is interpreted, acted on, and sustained over time. Without a healthy operating system, even the best metrics become noise. Design operations are the connective tissue that turns insight into action, turning information into alignment, and activity into momentum. It is where leadership meets infrastructure, and where care becomes something you can feel in the work.

Most people hear "design ops" and think about documentation, design systems, or process diagrams. Those things matter. But I believe operations go deeper. They shape how energy moves through a team, how decisions are made under pressure, and how people experience the work of designing together. Operations create clarity where there is fog, rhythm where there is chaos, and trust where there is uncertainty. In that sense, design ops is not just organizational work. It is culture work. It is how care becomes repeatable and how creativity scales without burning people out.

Now let's dig into the heartbeat of design: the invisible systems that keep teams healthy, hopeful, and in flow. It is where leadership and logistics meet. Where culture becomes infrastructure. Design operations live across the full arc of the organization, shaping how work moves from idea to execution, how teams collaborate across disciplines, and how creative energy is protected instead of drained.

Culture is not owned by a single role. Leaders are responsible for setting the conditions, but every designer participates in shaping the environment they work within. While this content draws from my experiences leading design teams, it is written for you, the designer. You do not need a title to influence culture. Through the rituals you support, the boundaries you ask for, and the care you extend to others, you help build the system you work inside. My hope is that something here, a practice, a question, a small operational shift, helps you become a stronger version of yourself while contributing to a healthier whole.

Culture Is a System We Build Together

Culture isn't the emoji reactions in Slack. It's not the team lunches or the all-hands playlist. Culture is what happens when things break. It's how trust is held, or dropped, when the pressure is on.

Culture shows up in a critique where someone shares an early idea, and the room stays curious instead of rushing to solve. It's in how leaders respond when someone burns out. It's in whether onboarding is seen as a checklist or a deeply human welcome.

People always ask how I scale design. But the real question was how I scale soul. In one of my earlier leadership roles, I joined a team that had gone through three massive reorgs in two years. People were weary. Design was seen as a service. Trust was low, and so was morale. I didn't start by fixing process. I started by tending to energy. I carved out space for team check-ins that were fully human. I made room for what-ifs. I made space for people to say what was actually on their minds. Designers do their best work when they trust that sharing a bold idea will be met with curiosity, not dismissal. When ideas are met with silence or indifference, creativity doesn't just stall, it slowly shuts down.

The turning point wasn't a new framework. It was a shift in how we related to one another. Slowly, culture started to take root. And only then did our systems start to work.

Culture is a mirror of what you tolerate, protect, and celebrate. I think it's important as a member of the team to understand what it feels like to get the support you need from your leadership. I believe that design leadership shapes design culture.

A Space for you to Thrive

The most impactful teams aren't just built on talent, but on trust, connection, and mutual respect. You deserve to work in an environment where collaboration is natural, communication is open, and inclusion isn't a side note; it's the starting point.

If you've ever worked in a place where design felt misunderstood or undervalued, you know how important it is to feel seen.

You're not just here to push pixels or make things look good, you're here to solve meaningful problems, connect ideas to people, and bring humanity into product experiences. That kind of work needs a home, a space where design is supported, protected, and empowered.

A design-friendly culture should feel like that: like home. It should offer space for deep thinking and heads-down focus. It should protect your time and your brain from endless context-switching, last-minute meetings, or unclear priorities. It's okay to ask for boundaries. Great design needs room to breathe.

And beyond just doing the work, you should feel supported in growing your craft. Whether it's through mentorship, conferences, learning a new tool, or exploring emerging tech, your curiosity matters. Staying sharp is part of the job, but staying inspired is what makes it sustainable.

You should feel safe sharing ideas, even the rough ones. You should be encouraged to take risks, learn from missteps, and shape the direction of the team. When your voice is heard and your ideas are valued, you're more than a contributor; you're a co-creator.

A healthy design culture isn't accidental. It's built with care, intention, and participation. You have a role in shaping it, just as much as anyone else. And when it's working, you'll know, because you won't just be doing good work. You'll be doing your best work, together.

Design Operations Is Leadership in Motion

Design operations is often misunderstood as backstage support work. But it's strategic. It's how workflows are shaped, how people grow, and how excellence becomes repeatable. If culture is the heartbeat, ops is the circulatory system, ensuring that clarity, care, and craft reach every part of the team.

Design ops should answer questions like:

- How do designers move through work?
- What are the tools the team is using, and are using them to the best of our ability?
- When a new team member joins, do they understand how we operate with ease, or is it confusing?
- How do we move from chaos to cadence?
- What rituals build team identity and trust?
- Where are we leaking energy? And how can we restore flow?

At organizations I have been part of, I saw how even small tweaks in our operating system freed up teams to spend less time in confusion and more time in creation. For instance, I reset our design review cadence or created a clear RACI for cross-functional decisions.

One of my favorite initiatives was introducing a "Monday Pulse", a 10-minute async check-in where designers reflected on their week ahead: what they were focused on, what they were stuck on, and one thing they were excited about. It seems small, but it was a cultural anchor. It gave the team a rhythm. And it gave me as a leader a live signal on who needed support, and where momentum was building.

Rituals Are the Soul in System Form

Every high-functioning design team I've been a part of has had one thing in common, meaningful rituals. Not performative ones, but rituals rooted in care and intention. These are the moments where culture becomes tangible.

A designer on one of my teams thanked me for setting the tone for cameras on and for people to get to know each other by a simple question everyone talked about their weekend and what they did. This creates motivation, it helps people be seen, and it helps designers feel safe.

Rituals aren't fluff. They're memory-making mechanisms. They create safety, rhythm, and identity. They help teams metabolize both wins and wounds.

Some of the most enduring rituals I've implemented or witnessed:

- **Critique with values.** Framing feedback through the lens of "How does this honor our principles?" versus "What's wrong with it?"
- **Design campfires.** Quarterly team sessions with no slides, just stories. We talked about what stretched us, what surprised us, what we learned.
- **Praise walls.** Where engineers, PMs, and other partners could shout out a designer, not for flashy work, but for meaningful collaboration.
- **"How we work" audits.** Once a quarter, team pauses and look at our rituals, our cadences, and our tools — asking what still served us and what needed to evolve. I was not in the room for those, as the team needed to have an open safe space and to be able to filter thoughts, feelings and their asks of me together, without me.

These rituals weren't optional. They were culture work. They reminded us that design is a human process, not just a production one.

Human Before Designer: Building Design Community

Design is not a solo sport. Even when you're deep in the quiet focus of pushing pixels or rethinking a flow, you're still designing *with* — with a team, with constraints, with a user you may never meet. That's why strong design communities don't just form by accident; they're cultivated.

Especially in remote or hybrid environments, building a sense of community takes intention. Cameras on. Monday rituals. Active listening. Real talk. Design reviews where you can be honest *and* kind. The spaces between the work matter as much as the work itself.

Design as a Social Practice

Design thrives in dialogue. Not every critique or standup needs to be perfect, but every interaction shapes how safe, connected, and energized the team feels. When designers have regular space to be seen, as humans, not just deliverables, they show up with more energy and honesty.

That's why I value things like Monday design calls. These aren't just status updates. They're grounding rituals where the team reconnects, aligns, and builds trust. Showing up with your camera on signals presence and respect. Hearing a teammate say "I struggled with this" or "I'm proud of this little detail" builds psychological safety in ways that polished decks never could.

A strong design community is where we can engage in healthy conflict too. Not everything will be easy, and that's a good thing. The goal is to normalize debate, not avoid it. Conflict in design, when framed with respect and shared goals, creates better outcomes.

Autonomy as a Trust Signal

Micromanagement kills creativity. But so does disappearing leadership. True autonomy is built through *earned* trust, and it goes both ways.

Sometimes I tell my team, "I'm going to get in the weeds here, not to take over, but to understand where you are." Designers need to know that deep questions are not a sign of mistrust, but of care. It's about co-thinking, not control.

I believe in activating thinking through inquiry: asking questions that open possibilities. What are we assuming here? What would happen if we flipped it? Is this solving the real problem? These kinds of prompts show that I trust my team's brains, not just their hands.

Autonomy also means having space to define your own working rhythms. It's okay to say, "I need focus time" or "Can we async this?"; as long as there's a shared understanding of outcomes and accountability. When autonomy is paired with clarity, it becomes a signal of mutual respect.

Motivation Through Connection

Designers are most energized when they feel connected: to the mission, to each other, and to the purpose behind the experience.

We all hit moments of doubt or fatigue. That's why community matters; it recharges motivation. Sometimes that looks like a Slack thread where people drop examples of great microcopy. Sometimes it's hopping on a call just to talk through a weird user edge case. Or celebrating when someone pushes a bold idea live, even if it wasn't perfect.

Design community gives people a reason to care; and to keep going. When people feel seen, included, and inspired, they don't just execute. They *create*.

Scaling Without Losing the Soul

Here's the paradox of scale: the bigger your team gets, the harder it is to feel connected. But the only way to scale *well* is to protect the connective tissue.

I've watched many orgs fall into the trap of prioritizing efficiency over empathy. They document everything, automate everything, and wonder why the team feels brittle. Scaling isn't just about hiring more designers or building better tooling. It's about preserving the things that made the team *feel* like a team in the first place.

To scale with soul, you need:

- **Codified values** that live in action, not just in notes.
- **Clear role expectations** that support growth, not hierarchy.
- **Shared language** around design excellence and impact.
- **Celebration rhythms** that acknowledge the *how*, not just the *what*.

One of the most meaningful exercises I ran with my team was called "What We Want to Be Known For." We wrote down, individually and collectively, the qualities we wanted to be recognized for by our partners. Thoughtfulness. Bravery. Listening. Craft. Then we mapped our rituals and systems to those traits. If our rituals didn't support those values, we rewired them.

Scaling soulfully requires this kind of ongoing inquiry. Otherwise, culture becomes accidental, and accidents tend to replicate the worst of what we didn't mean to build.

Operational Maturity

Building design culture needs to fit in context of the organization and design team and is not always linear, culture is very dynamic. For instance, a growing organization might have a startup-like innovation team experimenting with new ideas while its core product teams' function at an enterprise level. I'll outline characteristics of different stage organizations, from my experience, so you understand what you might need in your current organization.

Stage 1: Startup

- **Characteristics:** Small, scrappy, and highly dynamic. At this stage, the design team is usually a handful of individuals, if not a solo designer, wearing multiple hats.
- **Priorities:** The focus is on speed and impact. Processes are minimal, and the team is hyper-focused on delivering design work to meet immediate business needs. Collaboration tools might include sticky notes, quick Slack messages, and shared documentation.
- **Challenges:** Balancing the demand for fast turnarounds with the need for quality and consistency. The lack of formal processes can lead to burnout or inefficiency as the team juggles competing priorities.
- **Operational Focus:** Establish lightweight frameworks that make repetitive tasks smoother; think simple templates, design tool setups, or a shared folder structure.

Stage 2: Growth

- Characteristics: As the company scales, the design team grows too. This stage often sees the introduction of mid-level managers, more specialized roles (like UX researchers or motion designers), and a broader range of projects.

- Priorities: With more team members and stakeholders, coordination becomes essential. The focus shifts to building scalable processes, such as introducing a design system or creating guidelines for cross-functional collaboration.
- Challenges: Avoiding process overload. Teams in this stage often swing between chaotic creativity and overly rigid workflows as they experiment with finding the right balance.
- Operational Focus: Invest in tools and practices that scale, like project management platforms, onboarding resources, and formalized team rituals (e.g., design critiques). Begin tracking metrics like workload distribution or project timelines to monitor team health.

Stage 3: Enterprise

- Characteristics: In this stage, the design team is part of a large organization, often spread across multiple geographies or product lines. Specialized roles are the norm, and operations require a strategic and integrated approach.
- Priorities: The focus here is on harmonizing efforts across vast, complex structures. Teams need strong governance, mature design systems, and seamless integration with other functions like engineering, product, and marketing.
- Challenges: Maintaining a sense of creativity and connection in a large, distributed team. Bureaucracy and misalignment can stifle innovation if not managed carefully.
- Operational Focus: At this stage, formal operations become essential. Roles like Design Operations Managers or Program Managers focus on optimizing workflows, ensuring

budget alignment, and fostering a strong design culture. Metrics evolve to include team satisfaction, quality benchmarks, and ROI of design.

The Soul of Ops Is Human-Centered Leadership

At its best, design ops is an act of care. It's saying: we value your time, your creativity, and your growth — and we've designed our environment to reflect that.

Design leaders who embrace ops aren't just efficient. They're connective. They understand that culture is infrastructure. They notice where energy is leaking and where potential is blocked. And they design with, not just for, their teams.

I often think of ops not as a side function, but as the *exhale* that follows a deep creative breath. Without it, you're holding tension all the time.

Foundational Tools and Processes

Here are some key operation aspects for a design team that need to be addressed and decided on, some of these areas need more collaboration with other functions, like engineering or product, and some a design team can decide and strategize the right approach. Key questions to think about:

- Tools
    - What tools are you using, for discovery?
    - For knowledge repository?
    - For pattern discovery?
    - For getting close to customers?
    - For understanding the competitive landscape?

- Research
    - What types of research are you doing?
    - Are there partners in the organization that be utilized for more generative research activities?
    - What tools can you leverage for conducting the tests, for analyzing data, for recruiting users?
    - Where are findings stored?
- Process touchpoints
    - How do you collaborate with other designers (critiques, design pairing, partnering)?
    - Do you know who to go to for brand-related questions and helping your brand partners understand your work?
    - What kinds of touchpoints do you have with your product managers?
    - What kind of touchpoints do you have with engineers?

Carry This Forward

Culture is the system beneath the surface; the rhythm that sustains creativity. Carry forward the intention to nurture connection, protect energy, and build environments where you can thrive.

Chapter Reflection

Reflection Prompts

- What gives your design team energy right now? What drains it?

- What rituals, behaviors, or systems are shaping your team's culture, intentionally or not?
- How do you personally contribute to the emotional rhythm of your team?
- If your team had a heartbeat, what would it sound like? What would it need?

Artistic Expression

Draw your team's heartbeat. Not just a literal pulse, but a rhythmic pattern — how energy flows, pauses, syncs, or scatters. Add symbolic layers for rituals (candles, cycles, music), systems (wires, grids, circuits), and emotional connections (hands, eyes, breath).

Soulprint Words

Choose 3–5 words that capture what this chapter unlocked in you.

Examples: *Harmonizer, Culture Keeper, Energizer, Pulse-maker, Ritualist*

Soulprint Thread

This chapter reveals your role as a *keeper of creative culture.*

Carry forward a mark or rhythm that represents how you support others, co-create momentum, and breathe life into systems.

"I shape the invisible structures that help design thrive."

| 11 |

A Future Built on Belief

Every culture is built on belief. As we close, let's look toward the future of design; and the stories we choose to tell about what's possible.

Design is storytelling. Not the metaphorical kind, but the kind that helps people believe — believe in a product, a brand, a team, a vision, and most importantly, themselves.

Good stories change how people feel, act, and show up. That's what design does. We create artifacts that carry belief forward. It's easy to reduce design to systems, toolkits, and frameworks; but underneath all of it is the story. The why. The heartbeat.

As designers, we often talk about design thinking, customer journeys, value propositions, but we forget that we are crafting narrative arcs. We're helping people move from uncertainty to clarity. From friction to flow. From confusion to confidence.

We make people feel something. We build meaning.

Design is the Story People Tell Themselves

The experience of a product is not just what it does, but what it signals. How it makes someone feel about themselves. Great design makes someone feel smart, capable, included. It tells them, "You belong here. You've got this."

Design that works isn't always noticed; but the story it reinforces is. When someone glides through an interface and says, "That was easy" it's because the design told them a story: You're in control. You're welcome here. You're not lost. And that story stays with them.

That's why consistency in design matters; not just in visuals, but in tone, behavior, and emotional cues. That's how we make the experience trustworthy. Coherent design is coherent storytelling.

And when it's not coherent? People fill in the gaps with their own stories. If something feels off or janky or confusing, the story becomes: "This product isn't for me." "I don't get it." "This company doesn't care." That's what we're up against, not just usability issues, but narrative breakdowns.

Designers as Story Builders

Designers are not decorators. We are not just crafting how things look; we are shaping how they are understood. We're meaning-makers.

Every decision is narrative. That padding? It's a breath. That label? It's a tone shift. That empty state? It's the difference between feeling stuck and feeling invited. We're always telling a story, whether we realize it or not.

And that means we need to understand the full arc. Not just the happy path. What happens when something goes wrong? When someone cancels? When someone's confused? These are moments full of emotional weight, and full of opportunity. They're where belief is most fragile. Or most powerful.

Good storytelling isn't just about the beginning and the end. It's about the messy middle. The part where people doubt, question, and feel unsure. That's where design can offer a thread to hold onto.

Internal Stories Matter Too

It's not just what the product says to users, it's what the design says to the team building it.

Designers influence internal narrative just as much as external. The way we present work. The way we talk about tradeoffs. The way we advocate for the user *and* for the business. These are all acts of storytelling. And they shape how others perceive design's value.

If we always show pixel-perfect mocks without the mess, people think design is magic. If we never show the strategy, they think design is surface level. If we only speak in design language, they tune out. But if we connect the dots; user need, to business impact, to design decision, we become translators. We become trusted.

I've seen design teams change their standing in an org just by shifting how they tell their story. Not with more slides. But with better ones. Clearer ones. More human ones. That's the work.

Advocacy is a Belief System

Design is more than just making things look good. It's how we show up — for people, for teams, and for the world around us. At its core, design is advocacy. And advocacy is belief in action.

I've always held this close. It's not just a leadership philosophy, it's how I move through work. My team is always my first priority, but real advocacy doesn't stop at organizational lines. If someone in another function needs support, clarity, or a champion, I make space. Not because it's required, but because that's how healthy, high-performing environments are built, through attention, care, and shared ownership.

As designers, we wear many hats: problem-solver, creator, collaborator, one of our most powerful roles is that of advocate. We hold space. We speak up. We connect the dots. And we do it across three interconnected layers of belief: users, teams and impact.

We Advocate for Users

We are the voice in the room for the people who aren't in the room. That's our charge. We speak for the humans on the other side of the interface, especially when their needs are invisible, unpopular, or inconvenient. That means championing accessibility. Designing for inclusion. Asking who might be left out. Every choice we make, every pattern we reuse or question, should be grounded in empathy and intention. This is how we honor peoples lived realities, not just their clicks.

We Advocate for Teams

Design doesn't live in a vacuum. It lives in tension, with timelines, with competing priorities, with ambiguity and change. Advocacy, here, means creating clarity where there's chaos. It means standing up for quality, even when it's hard. Sometimes that looks like protecting your time so you can deliver on deadline. Other times, it means partnering with people in less-visible roles, legal, compliance, data security, and ensuring they're part of the story, too. Collaboration *is* advocacy. It brings dimension and richness to what we build.

We Advocate for Impact

We can't afford to only think about what we're building, we must ask why. What are the downstream effects of the choices we make? Are we contributing to ethical, sustainable systems, or perpetuating harm in ways we don't yet see? Design is a long game. It shapes habits, mindsets, and systems. Belief in better, more equitable, more thoughtful, more human outcomes, starts with us.

Empathy isn't a soft skill. It's a strategic one. It allows us to navigate ambiguity, work through conflict, influence across silos, and co-create with integrity. It's the lens through which we design, and the compass by which we lead.

Forget the myth of the lone genius. Your real power lies in activating others. In making space for the messy, collective, collaborative work of design.

Sometimes that looks like:

- Helping a PM sketch out a rough flow for a PRD
- Mocking up a quick UI for an experiment from marketing or Customer Success teams
- Writing documentation that helps customer support translate complex features
- Partnering with sales to tackle long-standing customer pain points
- Helping HR with a new review cycle process and system
- Designing a new process or ritual when teams need clarity, alignment, or a better way to move forward

These moments might seem small. But they build momentum. They build trust. They connect our craft to something bigger than the artifact. It opens the door up to broader thinking and understanding.

Design like you really mean it.

Advocate like you really mean it.

Because what you make — and how you show up to make it — can shift what people believe is possible.

Future of Design

Design has come a long way from its analog roots. What was once a solitary craft has become a collaborative, interdisciplinary, and data-driven practice. Designers today are not just makers; they are strategists, problem-solvers, and change agents. They operate at the intersection of creativity, technology, and business, shaping the way we live, work, and interact with the world.

Yet, as much as design has changed, some things remain constant. The core of design — the drive to create something meaningful, the commitment to understanding and serving people, and the pursuit of beauty and function — endures. The tools and contexts may evolve, but the heart of design remains the same. Our future is clear.

The future of design promises to be just as dynamic as its past. As we navigate the complexities of AI, globalization, and sustainability, designers will need to embrace adaptability, empathy, and curiosity. We will need to learn from our history while pushing the boundaries of what design can achieve.

Design will continue to reinvent itself. The question is not whether design will change, but how we as designers will rise to meet those changes. Will we cling to old ways of working, or will we embrace the unknown and lead the way into the future? The answer lies in our ability to dream, adapt, and create — just as it always has.

We're at a critical juncture where creativity feels endangered. Technology's rapid rise has brought us incredible tools but, in many ways, has dulled our curious and creative instincts. Designers are uniquely positioned to bridge the gap between technology and humanity.

I've even seen this pattern start early. In teaching kids' art, I've watched how quickly creativity narrows when children learn that approval comes from conformity. Many begin to believe that to be liked, successful, or "good," they need to make what's expected rather than what's imagined. Those same patterns follow us into adulthood and into our work.

It's too easy to lose sight of imagination when algorithms promise quick answers and AI handles the heavy lifting. But designers, we are the guardians of creativity. YES!!!! It's up to us to interrupt these learned habits and to unlearn the myths that hold us back. The myth that creativity is an elusive, magical force. The myth that designers are other-worldly artists disconnected from business realities. The myth that design is secondary to strategy when, in fact, it is strategy.

The world is demanding more from design than ever before. Design is no longer an afterthought or a veneer; it's the engine of innovation and the heart of user experience. The future calls for boldness, collaboration, and reinvention. Designers, this is our time to shape what's next.

We've Done This Before and We Will Do It Again

Design has never been static. It has always been an evolving discipline, adapting to the needs of the time and the technologies at hand. Over the past few decades, this evolution has been especially pronounced. From analog practices to the digital revolution, from the rise of user experience to the advent of artificial intelligence (AI), the design field has continuously reinvented itself. Let's explore this journey and what it reveals about the future.

The Human-Technology Partnership

As we navigate this transformation, let's remember that the essence of design remains the same: solving problems, creating value, and improving lives. What has changed is the scope of our influence and the tools at our disposal. By embracing these new responsibilities, designers can continue to lead with purpose and shape a future that is both innovative and human centered.

At its core, the partnership between humans and technology is one of co-creation. AI tools like generative design platforms, predictive algorithms, and automation systems are not merely extensions of human intent; they are active participants in the creative process. These tools can offer novel solutions, identify patterns that humans might overlook, and generate ideas at a scale and speed that would otherwise be impossible.

For designers, this co-creation requires a mindset shift. Rather than viewing AI as a competitor, we must embrace it as a collaborator. Consider how a painter might use a brush to execute their vision — the brush does not diminish the painter's creativity; it enables it. Similarly, AI can act as a medium through which designers explore new ideas, refine concepts, and challenge conventional thinking.

While AI excels at efficiency and precision, the essence of design lies in its humanity — the empathy, intuition, and narrative that connect people to the experiences we create. The challenge for designers is to strike a balance between leveraging automation and preserving artistry.

Take, for example, the role of AI in generating design variations. Tools like Adobe's Firefly or Figma's AI-powered features can create countless iterations of a layout in seconds. While this dramatically accelerates the process, it's the designer who must curate, interpret, and refine these outputs. It is the human touch that transforms raw possibilities into meaningful solutions. A human, just like AI, can output things that don't align. This goes back to intentionality. This is so so important, if you don't know why you are designing something and understand the value, the output from a human or AI will not work. Designers must critically assess how automation influences the user experience. Are we sacrificing nuance for speed? Are we ensuring accessibility and inclusivity? By asking these questions, designers can use AI not just as a tool but as a partner in crafting experiences that resonate deeply with more users.

I think one of the most exciting aspects of the human-technology partnership is its ability to push creative boundaries. I use AI to think beyond what I'm thinking about, to expand it. AI has the capacity to process vast datasets, simulate complex scenarios, and experiment with unconventional ideas. This capability allows designers to venture into unexplored territories and take risks that might otherwise feel insurmountable.

For instance, AI can analyze cultural trends to inspire designs that resonate globally while respecting local contexts. It can simulate user interactions to test concepts before they are even built, enabling a more iterative and informed design process. By harnessing these capabilities, designers can expand their creative horizons and deliver work that is both innovative and grounded.

The Role of Intuition

Despite its power, AI lacks one critical element: intuition. It cannot feel, imagine, or dream in the way humans can. This is where the designer's role becomes indispensable. While AI might suggest a thousand possibilities, it is the designer's intuition that discerns which path to follow. This interplay between human insight and technological capability is what elevates the partnership from functional to transformational.

Designers must trust their instincts while remaining open to the possibilities that AI presents. This requires a dynamic approach, where intuition guides the use of technology and technology, in turn, informs intuition. Together, they create a feedback loop that drives innovation and elevates the craft.

The human-technology partnership also fosters collaboration across disciplines. Designers often find themselves collaborating with engineers, data scientists, product managers and ethicists to shape the tools and systems they use. This interdisciplinary approach enriches the design process, bringing diverse perspectives and expertise to bear on complex challenges.

For example, when designing an AI-driven personalization system, a designer might work with data scientists to understand user behavior, engineers to implement the system, and ethicists to ensure it aligns with societal values. This collaborative environment not only enhances the quality of the work but also positions designers as leaders who bridge technical and humanistic domains.

As AI continues to evolve, so too will the nature of the human-technology partnership. Designers will increasingly find themselves in roles that require them to shape the behavior of AI systems, advocate for ethical standards, and explore the outer limits of creativity. This is both a challenge and an opportunity.

The future belongs to those who can embrace this partnership with curiosity and courage. By viewing AI not as a replacement but as a collaborator, designers can unlock new levels of creativity and impact. Together, humans and technology can create experiences that are not only efficient and innovative but also profoundly human.

The New Role of the Designer

The role of the designer is undergoing a profound transformation. I've seen this in my own reflection of my outputs and contributions over the past few decades. We are no longer confined to the creation of visuals or interfaces, designers are now stepping into roles that require them to be facilitators, coaches, strategists, and storytellers. I have and I implore designers to act in this way as well. This evolution reflects the changing demands of a world increasingly shaped by technology, data, and artificial intelligence. As the boundaries of design expand, so too does the responsibility and influence of those who practice it.

Traditionally, designers have been seen as creators — problem solvers tasked with delivering tangible outputs, whether it was a logo, a website, or a product. But today, designers are often called upon to facilitate conversations, align diverse perspectives, and co-create solutions with interdisciplinary teams. This shift requires a new set of skills: empathy, collaboration, and the ability to navigate complex organizational dynamics.

For example, in the context of AI-driven projects, designers might lead workshops to define ethical principles, ensure inclusivity, or align stakeholders around a shared vision. Our role as facilitators helps bridge the gap between technical capabilities and human needs, ensuring that AI systems are not just functional but also meaningful and equitable. I personally, have seen immense value in partnering with people leaders, with security officers, with lawyers, with data leaders because they all affect user experiences and I saw value in my impact to each of those functions and vice versa.

This interdisciplinary collaboration often requires designers to act as translators, making technical concepts accessible to non-technical stakeholders and vice versa.

Why This Book is a Story

The house was quiet. Dishes done, toys picked up, the hum of the dishwasher in the background. The kids were finally asleep. I should've gone to bed too. But instead, I opened my laptop.

Not to catch up on email or check Slack, those could wait. I just needed a minute to go back to something that had been sitting in the back of my mind all day. A decision we made earlier in a meeting that felt... off. Small, on the surface. But something about it tugged at me.

It wasn't about the layout or the flow. It was about the story we were telling — and the story we weren't.

So I pulled it back up. The wireframe. The copy. The deck. I'll do this for just for a few minutes, I told myself.

I didn't redesign the whole thing. I didn't need to. I adjusted a headline. Reordered a few points. Pulled forward the part that actually mattered but had gotten buried in all the business speak. Nothing flashy. Just... honest.

And then I closed the lid.

Sitting there in the dim light, I let myself feel it. That small, quiet spark that comes when something finally aligns. Not with best practices. But with *truth.*

That's what it means to design like you really mean it.

It's not about staying up late to prove something. It's about caring deeply enough to return to the work when no one's watching. It's listening to the nudge in your gut. It's standing up for the parts of the product that whisper, not shout.

Not everyone will notice.

But *someone* will feel it.

And sometimes, that's everything.

This book isn't a framework. It's not a set of rules or steps. It's a story. A collection of moments, lessons, and tensions I've lived and witnessed. It's not the only story of design. But it's mine. And maybe, parts of it are yours too.

I didn't write this to create another model. I wrote it because I believe stories are how we grow. We remember them. We share them. We repeat them to ourselves when we need courage or clarity. And we rewrite them, too, when we're ready.

This book is about the stories we hold of being both intuitive and analytical. Collaborative and independent. Strategic and craft driven. Ambitious and practical. These are not contradictions. They're the shape of a full, rich, human designer.

And if there's one thing I want to leave you with, it's this:

You are not just a designer.

You are a storyteller.

Every decision you make tells a story.

So make it a good one.

Carry This Forward

Belief is the thread that ties us together. It's the quiet conviction that design can change not just what people see, but what they believe. Carry this forward as both your compass and your call: design like you really mean it.

Chapter Reflection

Reflection Prompts

- What do you now believe about yourself as a designer that you didn't fully believe before?

- What kind of future do you want to help build, with your craft, your leadership, and your presence?
- Where will you need to stay brave to live out your belief?
- What will it look like for you to design like you really mean it, every day?

Artistic Expression

Create a visual "North Star." This isn't just your career goal; it's a symbol or scene that represents your belief in what design *can be.* Draw, paint, or collage it with light, movement, or metaphor. Let it be your reminder that the future isn't fixed, it's designed.

Soulprint Words

Examples: *Believer, Vision-holder, Future-builder, Aligned, Illuminator*

Soulprint Thread

We have now revealed your commitment; to yourself, to your craft, to the future you want to shape.

Carry forward your North Star.

"I believe in what design can become; and I'm part of making it real."

| 12 |

This Is What Design Is

Design is not what we make. It's how we move.

It's the bridge between intention and experience, between a business goal and a human heartbeat.

It's curiosity turned into clarity.

It's empathy translated into action.

It's the courage to listen longer, to stay open, to shape possibility even when the path isn't clear.

Design is a conversation, between people, between ideas, between what is and what could be. It lives in the tension between beauty and function, simplicity and complexity, soul and system.

Design is how we give form to care.

It's how we make meaning visible.

It's how we leave things better than we found them.

When you design like you really mean it, you don't just solve problems; you create possibilities. You don't just design products; you design belief.

This is what design is.

And this is what it means to be a designer.

| 13 |

Your Soulprint

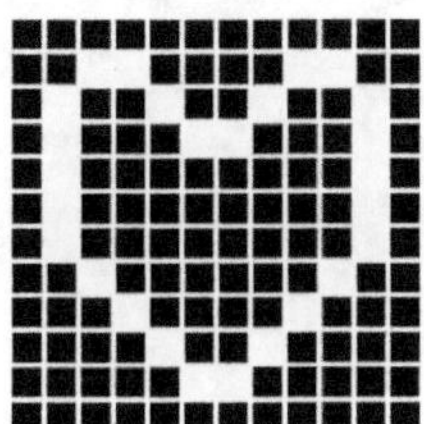

Soulprint Completion Prompt

You've done the work. You've listened, reflected, created. Now it's time to assemble your Soulprint.

Revisit your words. Your sketches. Your metaphors.

What patterns emerge? What energy flows across pages?

Now:

Choose a canvas: digital, physical, abstract.

Gather your chapter threads: your fire, your rhythm, your beliefs.

Bring them together as one visual, emotional, and aspirational artifact: **Your Designer Soulprint.**

It can be imperfect. Messy. Beautiful.

Let it remind you who you are when you design from your center.

You are not just a designer.

You are a system-shaper. A culture-keeper. A meaning-maker.

This is your future, *and you believe in it.*

About the Author

About the Author

Heidi Brown is a design executive, author, and Pacific Northwest native who believes design is at its best when it is intentional, humane, and deeply meant.

Based in Seattle, Heidi has spent more than two decades building and scaling product design organizations across fintech, SaaS, and consumer platforms. Her work lives at the intersection of product, systems, and craft, where strategy meets execution, and where design becomes connective tissue between business goals and human needs. She is known for strong design judgment, a clear point of view, and an ability to help teams scale without losing quality, creativity, or humanity.

Throughout her career, Heidi has led design in complex, regulated environments, partnering closely with product, engineering, marketing, and executive leadership. She has built design systems that support velocity without sacrificing craft, established accessibility and inclusive design practices in trust-critical products, and helped organizations mature design from execution to strategic influence. A consistent through-line in her work is mentorship: developing designers, leaders, and cross-functional partners by creating space for growth, feedback, and confidence in decision-making.

Outside of work, Heidi maintains a hands-on creative practice and a deep curiosity about the world around her. Hobbies include glass blowing, chainsaw carving, painting, photography, and dancing-studying tap and hip hop as forms of rhythm, discipline, and expression. She loves to travel and explore new places, drawn to how culture, environment, and movement shape perspective. An avid hiker, animal lover, and devoted hockey fan, she is especially inspired by pursuits that balance structure with instinct, and intensity with joy.

Heidi is also a mom of two girls, daughter, sister and wife, roles that ground her work in empathy, responsibility, and a long-term view of impact. She writes, speaks, and mentors with the belief that good design is not just about what we make, but how we help others grow.